ENGINES
OF
INNOVATION

ENGINES
OF
INNOVATION

U.S. Industrial Research
at the
End of an Era

Richard S. Rosenbloom
and
William J. Spencer

Editors

Harvard Business School Press
Boston, Massachusetts

Library of Congress Cataloging-in-Publication Data

Engines of innovation : U.S. industrial research at the end of an era /
   edited by Richard S. Rosenbloom and William J. Spencer.
      p. cm.
   Includes bibliographical references and index.
   ISBN 0-87584-675-0
   1. Research, Industrial—United States. 2. Technological
innovations—United States.   I. Rosenbloom, Richard S.
II. Spencer, William J.
   T176.E58 1996
   658.5′71—dc20   95-43852
                    CIP

Chapter 1 is adapted from "Industrial Research and Manufacturing Technology" by David A. Hounshell. Used by permission of Charles Scribner's Sons,
an imprint of Simon & Schuster Macmillan, from ENCYCLOPEDIA OF THE
UNITED STATES IN THE TWENTIETH CENTURY, Stanley I. Kutler, Editor
in Chief. Vol. II, pp. 831–857. Copyright © 1996 by Charles Scribner's Sons.

CONTENTS

PREFACE

IN 1993, a small group of experienced research managers and scholars of innovation met at a Harvard Business School conference to discuss the future of industrial research. Papers prepared for discussion at the conference led directly to the chapters in parts 1 and 3 of this book. Presentations made at the conference by executives gave rise to the chapters in part 2.

Both the conference and the preparation of this volume occurred under the joint auspices of the Division of Research of the Harvard Business School and of the inter-university Consortium on Competitiveness and Cooperation (CCC). Conceived by Professor David J. Teece and based at The Haas School of the University of California at Berkeley, the CCC promotes research, scholarly exchange, and dialogue with managers on a variety of topics related to U.S. competitiveness and the management of technology. Generous funding by The Alfred P. Sloan Foundation has contributed to its success. Our colleagues on the CCC Executive Committee provided continuing guidance and encouragement from the first suggestion of this project to its culmination in this book.

Thanks are due to many individuals who helped in various ways to bring this project to fruition. Valuable comments on one or more chapters were contributed by Harvey Brooks, Louis Galambos, Benjamin Gomes-Casseres, Margaret B. W. Graham, William Hamilton, Thomas MacAvoy, and Edward Roberts. The latter two were kind enough to review the entire manuscript. The ideas presented in the introduction and conclusion were first tested at the conference and were strengthened by the contributions of many of the participants. They were also presented at two meetings of the Directors of the Industrial Research in New York and at a meeting of the University Advisory Council of SEMATECH. We received encouragement and valuable critiques at these meetings, for which we are grateful.

Special thanks are due to Ralph Landau for wise counsel and assistance in the planning and execution of the conference and its follow-through; to Hirsh Cohen and Ralph Gomory for timely and helpful advice on various aspects of our work with the CCC; to John

Simon for outstanding editorial assistance before and after the conference and especially for his work on several chapters; to Jonathan West
for research assistance in preparation for the conference; to Rebecca
Henderson for an important contribution to the introduction; to Jackie
Baugher and her staff in Executive Education at Harvard Business
School for making things work so smoothly during the conference; to
Anne Englander for deciphering and transcribing numerous drafts; and
to Theresa Harrison upon whose exceptional administrative support
we relied throughout the project. We are grateful also to the Alfred P.
Sloan Foundation and to the Division of Research of the Harvard
Business School for the financial support without which this would not
have been possible.

ENGINES
OF
INNOVATION

INTRODUCTION: TECHNOLOGY'S VANISHING WELLSPRING

Richard S. Rosenbloom and William J. Spencer

THIS IS A BOOK about industrial research. It focuses specifically on research, rather than research *and* development (R&D), even though research accounts for a relatively small share of the more than $100 billion spent annually on industrial R&D in the United States.[1] Whereas most R&D is aimed at the improvement of existing products and processes or the development and commercialization of new products and processes, industrial research comprises more basic scientific investigation and the development of novel technologies. It expands the base of knowledge on which existing industries depend and generates new knowledge that leads to new technologies and the birth of new industries. Familiar exemplars of organizations devoted to industrial research include DuPont's Experimental Station in Wilmington, IBM's Watson Research Center in Yorktown Heights, AT&T's Bell Laboratories at Murray Hill, and the Xerox Palo Alto Research Center.

The decades following World War II were a golden era for research organizations in American industry (Graham 1985a). Corporate leaders, persuaded by the dazzling achievements of wartime and such undisputed research successes as nylon and the transistor, funded rapid expansion of research staffs and facilities. DuPont, for example, increased its R&D staff by 150 percent in the first decade after the war and the greatest growth occurred at the Experimental Station, the center of its fundamental research in chemistry. Radio Corporation of America (RCA), a pioneer in electronics, formed the RCA Laboratories Division, with an expansive budget and a campus-like setting in Princeton, New Jersey. Implicit in the rationale for these investments was a simple model of innovation: Scientific research in industry would generate a stream of inventions and discoveries that engineering would then make practical and affordable so that commercial organizations could harvest new revenues and profits.

1

The stream of industrial innovations that ensued from these investments fulfilled at least the first part of this ambitious corporate vision. Industrial laboratories proved to be wellsprings of important new technologies that then became staples of modern life, such as the integrated circuit, liquid crystals, and a multitude of new synthetic polymers, to name only a few. The interplay of university and corporate laboratories brought the development of lasers, techniques of recombinant DNA, and a host of computer technologies. Leading industrial laboratories gained stature as sources of important scientific knowledge. In the early 1960s, 70 percent of the papers appearing in *Physics Abstracts* on topics related to laser science were by authors working in industry (Bromberg 1991, 98). In the 1980s, as measured by citations to published papers, Xerox Corporation ranked on par with leading universities as one of the most influential sources of research in the physical sciences (see chapter 4).

Unfortunately, technological fecundity and scientific distinction did not always carry through to the corporate bottom line. Although the new science and technology created in industry contributed substantially to economic growth and productivity, corporate sponsors of the research often failed to capture significant returns. At first, buoyant demand in the U.S. economy in the 1950s and 1960s and the strong international market positions of the leaders in industrial research made it easy for those firms to sustain substantial investments in research even without clear evidence of immediate payoffs. The more competitive business environment of the 1980s, however, led managers to make more careful assessments of the profits gleaned from research.

As the business environment has become more competitive, the frontiers of industrial technology have shifted in ways that may make it more difficult for sponsoring firms to appropriate the profits inherent in the fruits of research. The richest opportunities now appear to be in fields different from those of a generation ago; today's frontiers are in biotechnology, exotic materials, and information sciences. The systems character of many electronic and information products (and services based on them) has forced firms to operate within broad cross-licensing webs that limit their opportunity to keep new technologies fully proprietary.

These competitive and technological trends have led to fundamental changes in the support of research by a host of firms. In their efforts to better understand how research can support their strategies for technology and innovation and how those strategies should fit within the operative and competitive behavior of their firms, corporate leaders have moved away from simple conceptions of the role of research in innovation and have transformed their research organizations. Research activities have been downsized, redirected, and restructured in recent years within most of the firms that once were among the largest sponsors of industrial research.

This volume will examine the character of the changes underway in the practice of research in industry. It will consider the likely consequences of these changes for industry and the economy as a whole and explore possible ways to mitigate potential adverse effects. We assert that these changes, overall, mark the end of an era for research in industry. As the ensuing chapters will make clear, it is easier to characterize the era now ending than to understand the nature of the era being shaped by the forces currently at work. Nevertheless, we hope that the papers assembled here will provide a basis for constructive consideration about the policy choices now emerging in industry, government, and higher education.

AN ECONOMIC PARADOX

As the twentieth century nears its end, the U.S. economy faces this paradox: While scientific research looms larger than ever as a stimulus to economic growth and a major component of American competitive advantage in the global marketplace, many leading American corporations are altering and shrinking the research organizations that should help to sustain that advantage.[2] Economic theory today emphasizes the stimulus to growth that is provided by the increasing returns characteristic of knowledge created by R&D (Grossman and Helpman 1991; Romer 1986). The very difficulty that firms have in capturing the benefits of new knowledge for themselves alone (that is, the tendency for knowledge to spill over to other firms), is now counted as a positive force for economic growth overall (Grossman and Helpman 1991). Furthermore, preliminary findings of empirical research suggest that there are high rates of return to research that is widely diffused across firms and industry (Griliches 1986; Jaffe 1989). There is accumulating evidence that there are important spillovers of knowledge across both firm and industry boundaries (Griliches, 1992; Jaffe 1986; and Jaffe, Trajtenberg, and Henderson 1993). To the extent that these findings hold for R&D broadly defined, they may be even more applicable to industrial research. The results of industrial research that is more "basic," or closer to "pure" science, are likely to give rise to a broad range of applications (increasing returns) and, thus, to find their way into the public domain (spillovers).

In the 1970s, and earlier, leading industrial sponsors of fundamental research, for example, AT&T, DuPont, IBM, Kodak, and Xerox, earned substantial portions of their revenues in markets in which they had shares of 80 percent and more. Profit margins comfortably covered the costs of research. In the 1990s, however, the globalization of markets, laissez-faire trade and regulatory policies in the United States, and the maturing of industries has forced these firms to face formidable rivals and narrowing margins in their core businesses. Each company has been forced by economic realities to reassess the scale, scope, and character of its research investments.

Furthermore, equity markets, which had valued the intangible technological assets of R&D intensive companies at a substantial multiple of their costs for decades, seem to have revised their valuations downward during the 1980s (Hall 1993). Economists now speculate whether or not there has been a significant decline in the returns to R&D (Griliches 1994). At least one academic feels no uncertainty about it; Michael Jensen (1993) has asserted that the R&D and capital expenditures of IBM, Kodak, and Xerox (among others) exceeded their value to shareholders by $5 to $10 billion each during the 1980s.

In response to the economic forces at play, the major enterprises in U.S. technology-based industries are beginning to reshape themselves—the market is working its will. One after another, these firms are restructuring, redirecting, and resizing their research organizations as part of a corporatewide emphasis on the timely and profitable commercialization of inventions combined with the rapid and continuing improvement of technologies in use. A recent authoritative survey of the American situation by the National Science Board (NSB) has concluded that "In large corporations, effort is shifting away from central [R&D] laboratories toward division-level effort with greater emphasis on risk minimization to meet the needs of today's customers" (National Science Board 1992, iii).

When the directors of Kodak decided to replace Kay Whitmore, the company's chief executive, they let reporters know that one of his most important failings was that he "spent too much on R&D without getting results" (Rigdon and Lublin 1993, 13). According to an article in the *Wall Street Journal*, investors applauded the move, urging Kodak to "become an 'aggressive follower' by capitalizing on rivals' inventions instead of mostly developing its own" (Cowan 1993). Reporters did not record which rivals were expected to replace Kodak as the technological pioneers that the company would imitate. IBM also cut the annual budget of its research division from $650 million to $500 million and redirected its focus toward "Services, Applications, and Solutions" [SAS] and away from investments in basic physical sciences and technology. The company planned to increase the SAS budget from 5 percent to 20 percent of the division's total between 1992 and 1994 with correspondingly sharp reductions in funding for basic sciences (Geppert 1994).

In addition to these changes among the largest and most mature research organizations, it is possible to observe other sorts of developments in industrial research. For example, more and more corporations operating on a global scale are establishing research centers outside their home countries. Japanese firms are opening laboratories in American centers of science, while U.S.-based firms are establishing laboratories overseas.

Some seasoned companies have continued to support pioneering research. Corning, for example, continues to concentrate most of its substantial R&D effort (more than 5 percent of sales) at its central laboratories and to focus on work aimed at "big hits" like CELCOR for catalytic converters (Morone 1993). Hewlett-Packard has also intensified its research commitment in the 1990s (see chapter 8). In contrast, however, most of the new breed of high-technology firms in electronics and information technologies—Apple, Intel, and Sun Microsystems, for example—have eschewed traditional research organizations and chosen other strategies.

Although important industrial research laboratories are being downsized and redirected, there does not seem to be a corresponding buildup in federal funds for basic and applied research in the nation's universities. A recent study by the National Academy of Science (1993) suggested further federal funding to support government and industry partnerships for identifying and developing technologies essential to national needs. While this is essential, given the present drive to reduce the deficit such funding will further reduce the funds available for long-term research, particularly in universities. There is some possibility that industrial funding will fill part of the increased need for R&D in universities. Through the Semiconductor Research Corporation, the semiconductor industry currently funds about $30 million per year of university research in semiconductor technology.

Hence, while industry seems to be moving its R&D dollars out of long-term basic research and into applied research, the same may be occurring within the federal government. This shift seems to have originated in Congress as a result of concern by legislators that the United States is falling further behind in technology development for critical industries, especially those industries in which there is a potential for rapid job growth.

The effects of these developments on industrial research are sure to be profound, although how they will affect it is still being debated. Edward David, presidential science adviser and former head of research at Exxon Corporation, flatly predicted that central laboratories would be eliminated (David 1994). On the other hand, John Seely Brown (1991) has argued that research "can reinvent the corporation," becoming central to the solution to today's competitive dilemmas. Either way, there are sure to be fundamental changes in the character of the industrial research laboratory as an institution in American industry and in the role of industrial research in the American economy. The chapters that follow are intended to provide an analytic and factual base for assessing the likely character and significance of these changes.

OUTLINE OF THE BOOK

This book is divided into three parts, followed by a brief, concluding chapter. In part 1, scholars of the history and economics of technological change analyze the institutional context of industrial research and innovation. In chapter 1, David Hounshell provides an historical overview of the evolution of industrial research in the United States during the twentieth century. He describes how the industrial laboratory emerged in this century as a distinctive scientific institution that was established in response to the private needs of corporate sponsors but shaped by the milieu of academic science and government funding in which it operated. In the 1990s, the end of the cold war, a new global competitive environment, and increasing fiscal constraints on public funds have reshaped the landscape of industrial research. An era has ended, but what is to come is still uncertain.

In chapter 2, Nathan Rosenberg and Richard Nelson study the interaction between industrial research and academic science and engineering. The roots of this interaction are deep, reaching well into the nineteenth century. On the eve of the twenty-first century, as Rosenberg and Nelson point out, both academia and industry must adapt to changes in the stance of government, stemming in part from the end of the cold war. The extent to which universities and corporations can help meet each other's needs in this new era remains to be seen.

In chapter 3, David Mowery and David Teece examine a relatively recent development in industrial research, the greatly enhanced significance of collaborative technical ventures that are undertaken as joint ventures, consortia, and university-industry alliances. They trace the forces that have given rise to this phenomenon, assess its prospects, and consider its implications for the performance of research in industry.

In part 2, the discussion switches from the general to the particular and from the observations of academic observers to the firsthand accounts of industrial executives. In chapters 4 through 6, top technical managers at Alcoa, IBM, and Xerox describe the changes that have occurred in the practice of research in their own firms, which have been among the leading supporters of research in industry. In the final chapter in this part, Gordon Moore describes why Intel, an undoubted leader in technology for the information age, elected not to establish a corporate research laboratory.

The accounts given in this part portray different attempts to redesign institutions and research practices in industry in order to reshape their role in the process of innovation. From the beginning, industrial research has had to address two distinct and sometimes conflicting goals: the creation of new science and technology; and the facilitation of the process of innovation, which translates the creative products of research into commercial results. One mark of the new era that is emerging is an increased demand for accountability in the achievement of this second goal, commercialization. Each of the two chapters in part 3 address this growing demand for commercial results. Both chapters combine the perspectives of academics who study innovation with those of executives who accomplish it. In chapter 8, Dorothy Leonard-Barton and John Doyle draw on the rich experience of Hewlett-Packard in commercializing technology. Chapter 9 builds on recent experience within Xerox as Mark Myers and Richard Rosenbloom argue for a "total process" view of the role of research.

The final chapter by Nelson, Rosenbloom, and Spencer explores the choices that industry, government, and universities face. The old assumptions about the proper conduct of research in industry will not hold in the era that lies ahead. The authors explore the dilemmas arising from this circumstance, finding no clear path toward resolving them.

Notes

1. Accurate estimates of the share of R&D accounted for by industrial research are difficult to obtain. Roughly 40 percent of government R&D funding is spent on basic and applied research, but the proportion spent in industry is certainly much smaller. Anecdotal evidence suggests that major industrial firms, such as Xerox, IBM, and AT&T, devote less than 20 percent of their R&D dollars to research.

2. We are grateful to Rebecca Henderson for calling our attention to several of the economic ideas that follow. Any errors of fact or interpretation here are the responsibility of the authors.

References

Bromberg, Joan Lisa. 1991. *The Laser in America, 1950–1970.* Cambridge: MIT Press.

Brown, John Seely. 1991. "Research That Reinvents the Corporation." *Harvard Business Review* (January–February): 102–11.

Cowan, Alison Leigh. 1993. "Unclear Future Forced Board's Hand." *New York Times*, 7 August, 1, 37.

David, Edward R., Jr. 1994. "Science in the Post-Cold War Era." *The Bridge* 24 (Spring): 3–8.

Geppert, Linda. 1994. "Industrial R&D: The New Priorities." *IEEE Spectrum* 31 (September): 30–41.

Graham, Margaret B. W. 1985. "Corporate Research And Development: The Latest Transformation." *Technology in Society* 7: 179–95.

Griliches, Zvi. 1986. "Productivity, R&D and Basic Research at the Firm Level in the 1970s." *American Economic Review* 76, no. 1: 141–54.

———. 1992. "The Search for R&D Spillovers." *The Scandinavian Journal of Economics* 94: 29–47.

———. 1994. "Productivity, R&D and the Data Constraint." *American Economic Review* 84, no. 1: 1–23.

Grossman, Gene, and Elhanan Helpman. 1991. *Innovation and Growth in the Global Economy.* Cambridge, Mass.: MIT Press.

Hall, Bronwyn. 1993. "New Evidence on the Impacts of R&D." *Brookings Papers on Economic Activity: Microeconomics.* Washington, D.C.: Brookings Institute.

Jaffe, Adam. 1986. "Technological Opportunity and Spillovers of R&D." *American Economic Review* 76: 984–1001.

———. 1989. "Real Effects of Academic Research." *American Economic Review* 79, no. 5: 957–70.

Jaffe, Adam, Manuel Trajtenberg, and Rebecca Henderson. 1993. "Geographic Location of Knowledge Spillovers as Evidenced by Patent Citations." *Quarterly Journal of Economics* (August): 577–98.

Jensen, Michael C. 1993. "The Modern Industrial Revolution, Exit, and the Failure of Internal Control Systems." *Journal of Finance* 48, no. 3: 831–80.

Morone, Joseph G. 1993. *Winning in High-Tech Markets: The Role of General Management*. Boston: Harvard Business School Press.

National Academy of Sciences, Committee on Science, Engineering, and Public Policy. 1993. *Science, Technology, and the Federal Government: National Goals for a New Era*. Washington, D.C.: National Academy of Sciences.

National Science Board. National Science Foundation. Committee on Industrial Support for R&D. 1992. *The Competitive Strength of U.S. Industrial Science and Technology: Strategic Issues*.

Rigdon, Joan E., and Joann S. Lublin. 1993. "Kodak Seeks Outsider to Be Chairman, CEO." *Wall Street Journal*, 9 August.

Romer, Paul. 1986. "Increasing Returns and Long-Run Growth." *Journal of Political Economy* 94, no. 5: 1002–37.

PART ONE

THE INSTITUTIONAL CONTEXT OF RESEARCH AND INNOVATION IN INDUSTRY

CHAPTER 1

The Evolution of Industrial Research in the United States

David A. Hounshell

CHAPTER 2

The Roles of Universities in the Advance of Industrial Technology

Nathan Rosenberg and Richard R. Nelson

CHAPTER 3

Strategic Alliances and Industrial Research

David C. Mowery and David J. Teece

1

THE EVOLUTION OF INDUSTRIAL RESEARCH IN THE UNITED STATES

David A. Hounshell

FORMALLY ORGANIZED INDUSTRIAL research and development (R&D) in the United States has existed for approximately a century. Depending upon how industrial R&D is defined, its first appearance lies somewhere between 1875, when the Pennsylvania Railroad, one of the nation's first "big businesses," hired a doctor of chemistry to bring science to bear on practice, and 1900, when General Electric's management established a laboratory devoted to generating and using scientific knowledge with the hope that this knowledge would keep the company's core technologies from being undermined by individuals and other companies. The faith in the utility of science displayed by those executives of the Pennsylvania Railroad and General Electric was not fundamentally different from the faith displayed by such early members of the American Philosophical Society, "held at Philadelphia, for Promoting Useful Knowledge," as Benjamin Franklin, David Rittenhouse, and Thomas Jefferson.[1] Even today, most people do not question the utility of knowledge, yet the world of Franklin, Rittenhouse, and Jefferson and the world today are fundamentally different. The pursuit of science is no longer only the avocation of learned, enlightened gentlemen but the activity of highly educated, specialized, and professionalized men and women who work in universities, institutes, and corporations. This dramatic transformation in the pursuit of science has been accompanied by equally dramatic transformations in business, government, the military, and technology. From a confederation of small farmers, craftsmen, and merchants who operated in a predominantly mercantilist world economy, the United States has become an industrial nation in which some corporations have sales that exceed the gross national product of most of the world's other nations.

Without question, technological change drove these transformations, and in the twentieth century, industrial R&D has occupied a critical role in this technological change. Indeed, as we approach the end of the twentieth century, the development of industrial R&D must surely rank as one of the century's defining characteristics. Of course, definitions of industrial R&D vary greatly. In 1948, the editor of a volume prepared for the Industrial Research Institute wrote that industrial "research, like poetry, cannot be defined in a manner that is universally acceptable" (Furnas 1948, 2). In 1959, Richard R. Nelson, one of the first economists to devote serious attention to industrial R&D, implicitly defined it by linking it with the functions of research and development laboratories:

> Research laboratories may be created and maintained by firms for many purposes, including development and application of quality control and other testing techniques, elimination of manufacturing troubles and improvement of manufacturing methods, improvement of existing products and development of new uses for them, development of new products and processes, and scientific research to acquire knowledge enabling more effective work to be done to achieve the above purposes. (Nelson 1959, 119)

More recently, Leonard Reich has characterized industrial research as research that is performed in "industrial laboratories set apart from production facilities, staffed by people trained in science and advanced engineering who work toward deeper understandings of corporate-related science and technology, and who are organized and administered to keep them somewhat insulated from immediate demands yet responsive to long-term company needs" (Reich 1985, 3). For accounting purposes, the nation's custodian of data on scientific research, the National Science Foundation (NSF), has developed formal definitions of basic research, applied research, and development. Although these definitions encompass industrial R&D, they are by no means precise and do not necessarily conform to accounting definitions used by firms that engage in extensive industrial R&D (NSF 1993, 94).

According to NSF, in 1993 the United States spent roughly $161 billion on R&D, or about 2.6 percent of the country's total gross domestic product (GDP). Of these expenditures, industry provided about 52 percent of the funds, while the federal government supplied 42 percent. Colleges and universities, state and local governments, and nonprofit institutions funded the remaining 6 percent. The federal government spent 46 percent of its share of the national R&D budget in industry and affiliated federally funded research and development centers. The remainder was spent on in-house governmental laboratories, colleges and universities, and nonprofit organizations. Thus in 1993, industry conducted slightly more that 68 percent of the nation's R&D on a dollar basis, or 1.77 percent of total GDP. The majority of all formally educated scientists and engineers in the United States—almost two million in 1992—worked in industry. Of these, a smaller number (950,000 in 1989) worked on a full-time basis in R&D. Almost 75 percent of these individuals worked in industry, approximately 18 percent in colleges and universities, and roughly 6 percent in federal agencies (NSF 1993).

This essay will focus primarily on the "R" of industrial R&D. Research usually represents the most problematic aspect of R&D and certainly is the minority partner when evaluated in terms of dollar expenditures. In 1991, about 28 percent of all industrial R&D expenditures were devoted to basic and applied research as defined by the NSF. The remaining 72 percent went to development. Percentages like these have been typical of much of the post–World War II era. Development has consumed a large part of most industrial R&D budgets because it entails the cost of making and testing prototype products or designing, constructing, and operating production equipment larger than the equipment usually used in the laboratory but smaller than commercial manufacturing equipment in order to perfect manufacturing processes for a new product or to test new processes for making an existing product. This work is expensive compared to the cost of the discovery, advancement, and application of new knowledge, which is the central concern of research. In the absence of research, however, most people believe there would be no development or, at least, a very different type of development. Yet, because the outcomes of research are harder to measure and predict than those of development, research is perceived to be a riskier undertaking.

This essay maps the principal factors that have given rise to industrial research, analyzes various forces that have mediated its spread, identifies some of the major actors in its history, and suggests both its significance and its significant trends. Although shaped by many factors both internal and external to industry, industrial research itself has been an important shaper of the twentieth century.

THE PREHISTORY OF INDUSTRIAL R&D

Historians seeking the roots of twentieth-century American industrial research have found scattered but numerous instances in the early nineteenth century of scientists contributing to the development of new technologies and industries.[2] Although interesting, these individual instances do not create a discernable pattern of science applied to industry until the 1870s when the telecommunications, electrical, and chemical industries, or "science-based" industries as historians have labeled them, began to emerge.

The increasing application of science in American industry was accompanied by the professionalization of American science. Science became an occupation devoted to the generation of new knowledge about the physical and biological worlds. Ironically, this professionalization was matched by ever more strident calls for the "independence" of science. Pleas poured forth for "pure" science, that is, science pursued for its own sake, and "pure scientists" as opposed to what some called "prostituted" science, science pursued for profit.[3] The shrill tone and bitterness of Henry A. Rowland, a distinguished physicist at Johns Hopkins University, as he addressed the American Association for the Advancement of Science in 1883 is indicative of the strength of feeling among some scientists at the time.

> The proper course of one in my position is to consider what must be done to create a science of physics in this country, rather than to call telegraphs, electric lights, and such conveniences, by the name of science. . . . When the average tone of the [scientific] society is low, when the highest honors are given to the mediocre, when third-class men are held up as examples, and when trifling inventions are magnified into scientific discoveries, then the influence of such societies is prejudicial. (Rowland 1902, 609)

While echoing earlier calls made by other major figures in the emerging scientific community, Rowland's widely circulated address also foreshadowed the later, more strident calls for "pure" science that would be heard.[4] Indeed, the pursuit of "pure" science, or "best science," forms one of the most important motifs in the history of American science and continues today to shape American science and the course of industrial research.[5]

Rowland and other American scientists (many with Ph.D.'s from German universities), who were beginning to populate what would later be termed "research universities," found it galling that the American public confused such inventions as the telegraph and the electric light bulb with science and mistook the inventors of those devices for true scientists.[6] Even more galling was how successful independent inventors were at garnering resources while scientists were forced to beg for support from an American public that did not appreciate the beauty of "pure" science (Hounshell 1980, 612–17). Not only did the American public confuse science with invention, it also believed that the true wellspring of invention and innovation was the heroic individual inventor. Even many of the leaders of technologically based industries and firms (for example, the Pennsylvania Railroad) did not appreciate the role of the scientist.[7]

When the managers of the Pennsylvania Railroad hired Charles B. Dudley in 1876 to establish a chemical laboratory in Altoona, Pennsylvania—perhaps the first corporate laboratory in the United States—they did not intend to shift the locus or nature of technological innovation and competition in the railroad industry. By the 1870s, those patterns of innovation and competition in the industry were well entrenched. Given the expense and complexity of railroad operations, technological innovation within the Pennsylvania Railroad, for example, was largely incremental and not the province of any single individual or department. Many technological innovations were carried out in the locomotive and car shops in Altoona, where the Pennsylvania exercised considerable leadership in design and construction. Still, many innovations were highly evolutionary in character, because of the extensive system into which they had to fit (Usselman 1985). More radical change came from outside, from independent inventors and engineers such as George Westinghouse. His invention of the air brake and, later, of important new switching and signaling devices and his shrewd patent management gave him a proprietary position that even the railroad industry could not break. While Westinghouse maintained a strong, proprietary, independent position in the industry, few other independent inventors and entrepreneurs duplicated his performance (Usselman 1992). For the most part, independent inventors sold their patented improvements and then moved on to other things.

Thus, although a recognized leader in locomotive development and overall system efficiency through the adoption of the latest innovations in rails, switching, and signaling, the Pennsylvania had never sought to develop proprietary technologies that could provide a competitive advantage. Instead, the Pennsylvania and other railroads used patents only to protect themselves from outsiders—independent inventors who might attempt to extract high rents through proprietary inventions.[8] Innovations by any large railroad quickly diffused to the others through a wide array of networks ranging from locomotive manufacturers to visiting delegations of engineers, industrywide standard-setting committees, patent pools, and industry publications.

Dudley's laboratory, therefore, focused on the standardization and testing of supplies required by the railroad: steel rails, lubricating oils, and so on. Dudley devoted his entire professional life to the establishment of standards for industrial materials. His work at the Pennsylvania developed into a national movement that was institutionalized as the American Society for Testing Materials, with Dudley as its longtime president (Usselman 1985).

The telegraph industry, which was also system driven, mirrored the railroad industry's pattern of innovation for most of the nineteenth century. This is not surprising, given that the telegraph and railroad industries grew hand-in-hand and were, as Chandler has argued, highly dependent on one another (Chandler 1977, 89). Established by Samuel F. B. Morse and his associates in the late 1840s, the telegraph industry faced many of the same imperatives as the railroad, with the only difference that the telegraph moved information rather than freight and passengers. Developing instruments to transmit and receive messages faster, simultaneously, and with greater reliability became the order of the day. In his study of technological innovation in the American telegraph industry, Paul Israel (1992) has argued that the machine shop, rather than the laboratory, was the locus of technological change in the telegraph industry for most of the nineteenth century. These shops were owned and staffed by mechanic-inventors who had begun their careers as telegraph operators and who often made inventions in response to specific requests by telegraph system superintendents. These mechanic-inventors brought practical skills and intimate knowledge of the industry's technology, rather than theoretical scientific training, to the process of invention. Indeed, their work was almost entirely devoid of theoretical development. Although the dominant firm in the industry, Western Union, eventually developed an organizationally distinct group of operator-inventors to integrate new technologies into Western Union's system, for a long time, the company relied on individual inventors, including Thomas A. Edison, for technological innovation. In fact, the company would not have a laboratory comparable to some other corporate labs until the World War I era (Israel 1992).

Edison worked at inventing for the telegraph industry in various machine shops in Boston, New York, and Newark, New Jersey. He was a partner in several of these shops, and in some of them, he equipped separate areas that he called laboratories. In 1876, he established his own laboratory/machine shop which he referred to as his "invention factory," or, more often, his laboratory, in the rural hamlet of Menlo

Park, New Jersey. Although some have seen the Menlo Park laboratory as the forerunner of the organized industrial research and development laboratories of the twentieth century (Albert North Whitehead [1985, 120] once said that the "greatest invention of the nineteenth century was the invention of the method of invention"), Edison founded his own laboratory to invent for a distinct market: the telegraph industry.[9] Inventors like Edison—and there were several—found a market for their inventions and sold their patented improvements to the highest bidder. The Menlo Park operation seems similar to later industrial research laboratories principally because it was there that Edison and his growing band of machinists, glassblowers, instrument makers, chemists, and even a physicist or two founded the nation's first commercial system of incandescent lighting.[10] Since, however, Edison abandoned the idea of inventing for an existing market and focused on building a new industry after 1878, Menlo Park could be viewed equally well as one of the nation's first high-technology start-up companies (Hounshell 1989, 116–33). Succeeding in this, Edison closed the Menlo Park complex in 1882 and became an industrialist. As such, he never developed the idea of building a research function geared to the generation and application of new scientific and technological knowledge. Only in 1887 did he realize that invention, not capitalism, was his forte. He then built a much larger and more thoroughly equipped laboratory in West Orange, New Jersey, and returned to the business of inventing.[11]

By that time, Edison's name was a household word, and the attention accorded to him and his new laboratory was deeply disturbing to scientists such as Rowland. The public's continuing confusion of invention with science and inventors with scientists strengthened the resolve of research-oriented scientists to impart the highest scientific ideals to the growing ranks of graduate students. For these scientists, the German system offered a far better model of how knowledge should be pursued, supported, and applied.

THE GERMAN MODEL OF SCIENTIFIC INDUSTRIAL R&D

During the last third of the nineteenth century and the first decade and a half of the twentieth century, a preponderance of U.S. physicists and chemists earned graduate degrees in Germany. These scientists not only observed a different system of graduate education in German universities but also witnessed the evolution of a nationwide system of scientific research and industrial-academic relations that had begun after German unification in 1871. By the beginning of World War I, Germany possessed the world's most complex and advanced research system, comprised of university research programs, government- and industry-sponsored research institutes, and industrial R&D programs.[12] These components were linked by research scientists committed to advancing science and, as necessary to this goal, technology and industry.

A key element of the German system was industrial sponsorship of research in universities. The directors of German firms, especially in the chemical and electrical industries, believed that the interests of their firms and those of professors were mutual.[13] Thus, firms would invest, sometimes on an exclusive basis, in individual professors and their cadres of graduate students. For their investments, the firms gained access to new scientific knowledge and were able to recruit high-caliber graduates for the product development and testing laboratories they were organizing and expanding. The professors gained access to expensive or impossible-to-get industrial materials, chemicals, and instruments and a window on evolving industrial practice. Created through private philanthropy and state action and supported in part by industrial funds, German research institutes generated basic knowledge that was important to university professors and industrial scientists alike.[14]

In the two decades before the turn of the century, German firms in the newly created electrical and chemical industries gradually built up in-house R&D organizations that eventually became the models for all science-based industries. In the chemical industry, the experience of the firm Friedrich Bayer, A.G., is representative. The firm tried several methods for harnessing science and scientists to its commercial objectives, and by 1891, the major patterns and forms of its R&D programs had crystallized under the leadership of Carl Duisberg, the firm's first research director who would later become a board member, then president of Bayer, and eventually the creator of the large chemical combine, I. G. Farben. Bayer built a central research laboratory furnished with the latest scientific instruments and supplied with a scientific and patent library and a seminar room. The company also established specialized applications laboratories and other facilities dedicated to improving production processes. Scientists, most of whom had earned doctorates from German research universities, staffed all of these facilities.[15]

Henry Rowland and other American scientists admired the German research system because of the level of scientific research in the universities and institutes that drove the country's system of graduate education. At the turn of the century, research-based graduate education was only beginning to develop in the United States. American managers of technologically advanced firms admired the German system because of the evident success of the industrial R&D programs in the German chemical and electrical industries. Until 1900, no U.S. firm had an R&D organization comparable to those of such German firms as Bayer, BASF, and Hoechst in the chemical industry and Siemens in the electrical industry. Despite the hopes and plans of many American scientists, nothing comparable to the research system of imperial Germany existed in the United States at the turn of the century.

The R&D Pioneers

Eventually, a research system that was at least comparable to the German system did evolve in the United States. This system took root in the first two decades of the twentieth century and was conditioned by many forces, some of which were unique to the United States. The handful of firms that pioneered industrial research programs included General Electric (GE), American Telephone & Telegraph Co. (AT&T), E. I. du Pont de Nemours & Co. (DuPont), and Eastman Kodak (Kodak), as well as a few other companies, including General Chemical (laboratory founded in 1900), Dow (1900), Standard Oil of Indiana (1906), Goodyear (1909), and American Cyanamid (1912). Among these other five companies, only Standard Oil's research exerted considerable influence on national patterns of R&D in the period before World War I.[16]

The founding of formal R&D programs by these manufacturers stemmed in part from competitive threats to their businesses or core technologies. General Electric, for example, which arose from the 1892 merger of Edison General Electric and the Thomson-Houston Company, established its General Electric Research Laboratory (GERL) in 1900 in large part because its managers feared that the firm's incandescent lamp business would be rendered obsolete by new lighting technologies.[17] Even without technological threats, which included radical improvements to gas lighting, gas-filled electric lights being developed by German academic and industrial scientists, and a promising electric light under development by the New York inventor Peter Cooper, GE faced the entry of new players into the highly profitable electric lamp business because Edison's basic patents were expiring. Believing that GE was ill-equipped to deal with chemical matters and thus might be overwhelmed, Charles Steinmetz, GE's brilliant chief consulting engineer, convinced his superiors to establish a laboratory that would gather, advance, and apply the kind of chemical and electrochemical knowledge that seemed necessary to check or outdo the competition.[18]

Steinmetz, Elihu Thomson, then an active consultant to GE, and Edwin Rice, GE's vice president who was responsible for engineering and manufacturing, recruited Willis R. Whitney, a physical chemist from the Massachusetts Institute of Technology to be the founding director of GERL. Because Whitney, who had done his Ph.D. work in Leipzig under Wilhelm Ostwald, at first would not consider leaving the academic world completely, Rice worked out an agreement whereby Whitney would commute from Boston to Schenectady two days a week to direct research at GERL.[19] Whitney was by no means opposed to the use of science for industrial or commercial purposes; like others he had derived industrial income from his intellectual labors, but working full time for an industrial company did not fit the "pure" science ideology to which he subscribed.[20] Arthur D. Little once reported that in the 1890s Whitney had spurned his offer of a position with Little's consulting business at twice his MIT salary, telling Little he would rather be an academic "than be president" (Wise 1985b, 62). Thus, from the beginning, GERL was shaped by the "pure" science ideology of the time.

Indeed, although Whitney soon recognized that the GE job offered intellectual challenges every bit as exciting as those of his MIT position, he obviously worried about his status within the academic community. Many of the decisions he made during his career at GE were driven as much by these concerns as they were by the company's business concerns. Consequently, even the mathematically gifted, fiercely opinionated Steinmetz soon began to criticize the scientist's approach to industrial research. Believing Whitney to be too focused on building a scientific research organization, Steinmetz confided that the GERL was not "what I hoped for it" (Kline 1992, 150). Instead of backing Whitney, he set up his own "creative engineering" organization within General Electric to do the things he thought Whitney should be doing (Kline 1992, 149–51; Brittain 1976).

In fact, Whitney had to deal with the ideology of "pure" science every time he wanted to recruit a physicist or chemist.[21] Therefore, he had the difficult, dual task of proving to his bosses that investment in science could pay off and to his academic peers and recruits that GE's research program could compete with those of the new research universities. This burden was carried by all the research directors of the R&D pioneers, but Whitney succeeded far better than most. The GERL distinguished itself through its development of ductile tungsten filaments in 1910 and then the gas-filled light bulb in 1913, which was far more efficient than Edison's carbon-filament lamps and outclassed the metal filament lamps developed in Europe. The lamps gave the company a proprietary position in the electric lamp industry and an important leg up in development of radio vacuum tubes. GERL's work on gas-filled lamps also yielded the lab's first Nobel Prize in 1932 in recognition of Irving Langmuir's surface chemistry research.[22] For Whitney, however, the price of the lab's success was high. He suffered nervous breakdowns three times during his long tenure at GE, the first one six years after he had joined GE full time and before the lab had begun to produce either scientifically outstanding or financially rewarding work. Whitney's last breakdown occurred when the Great Depression forced him to fire a large number of scientists, tarnishing the laboratory's and his own reputation and ending his career.[23]

Competition or fear of competition also drove the establishment of formal R&D programs in the other R&D pioneers. At first, AT&T followed the telegraph industry's practice of relying on the market for technological innovation. Over time, however, it began to develop technical talent in both its telephone division and its manufacturing arm, Western Electric. The expiration of the major Bell patents and the growth of a large number of independent telephone companies propelled the company toward organizing a concerted research and development function. By 1909, the company faced a new competitive threat—radio—which if not controlled might render its investment in wires obsolete. In response, AT&T launched a major research program, staffed by physicists (most of whom were trained in European graduate programs and some of whom did brilliant science) and theoretically inclined engineers. These researchers conducted work in electronics and communications and circuit theory with the goal of controlling radio. At the same time, AT&T adopted an aggressive strategy to build what Theodore Vail, the company's president, called "universal service," which ultimately translated into developing coast-to-coast telecommunications capabilities from anywhere in the system. Supported by Vail, AT&T's chief engineer and research director, J. J. Carty, committed the company to coast-to-coast long-distance service by 1915. The technical problems of controlling radio and developing coast-to-coast telephone transmission were intimately related. As Leonard Reich (1985) has shown, the R&D efforts coordinated by Carty and his assistant, Frank B. Jewett, a physicist, gave large-scale industrial R&D a permanent place in the Bell organization. By 1925, when Bell Laboratories was formally incorporated as a subsidiary of the company, the AT&T labs employed more than 3,600 staff members and operated on a budget in excess of $12 million, by far the largest R&D program in the United States.[24] (In 1925, DuPont spent slightly less than $2 million on R&D while GE devoted about $1.4 million [Hounshell and Smith 1988, 288; Wise 1985b, 246].)

The potential competitor for DuPont was the U.S. government.* During the 1890s, the navy had encouraged DuPont, which had long supplied black powder to the navy and army, to develop manufacturing capabilities and capacity in guncotton or smokeless powder (nitrocellulose). However, after developing these new capabilities, and especially after a new generation of the du Pont family began to consolidate roughly two-thirds of the American explosives industry after 1903, some members of the military and Congress decided that the government should not be dependent on a trust for powder or innovation in explosives.

DuPont established the General Experimental Laboratory, a forerunner to its Experimental Station, in 1903 to develop, among other things, improved products and processes for smokeless powder and a deeper understanding of nitrocellulose chemistry. A year earlier, the company's high explosives division had founded a research organization that quickly conveyed how successfully science could be harnessed to industry. Known as the Eastern Laboratory and directed by a German-trained chemist named Charles L. Reese, it was created primarily to gain better control over manufacturing processes, to develop safer high explosives that would meet anticipated regulations for "permissible" explosives, and to lower manufacturing costs. While Reese could easily demonstrate how a dollar invested in R&D at Eastern Laboratory yielded direct returns of three dollars of earnings through new products or improved manufacturing processes, Reese's counterpart at the General Experimental Laboratory was unable to show such direct profits. Nevertheless, considering the indirect benefits of DuPont's R&D, the General Experimental Laboratory's work on smokeless powder yielded staggeringly successful results. It proved to be the means by which the company retained all of its smokeless powder capacity (indeed, all the smokeless powder capacity in the United States) after

---

* The following description of DuPont's early research programs is based on David A. Hounshell and John Kenly Smith, Jr., *Science and Corporate Strategy: DuPont R&D, 1902–1980* (New York: Cambridge University Press, 1988), 11–110.

DuPont was found guilty of violating the Sherman Antitrust Act in 1912. Through the intercession of the military, which recognized that DuPont had devoted considerable sums to innovation in smokeless powder, the antitrust consent decree left this segment of DuPont's explosives business intact, just in time for the outbreak of World War I in Europe. (DuPont was, however, forced to divest two-thirds of each of its other two divisions—high explosives and black powder.)

Competition and fear of government antitrust action also drove George Eastman to found his highly productive laboratory at Kodak Park in Rochester, New York, in 1912. As the German synthetic dye industry matured, many German firms turned their researchers to the development of other fine chemicals, including photographic chemicals and film. Keeping ahead of the competition by continual product innovation had been Eastman's official strategy since at least 1896.[25] As Kodak's business continued to expand, however, it came to rely more and more heavily on the German chemical industry for its intermediates and photographic chemicals. Such reliance on firms that were aggressively building a strong international presence in Kodak's product lines posed a real threat to the firm's long-term health. While at a dinner on a business trip to Europe in late 1911 or early 1912, Eastman sat next to Carl Duisberg, then head of Friedrich Bayer, A.G. Duisberg casually told Eastman that several hundred chemists with doctorates were employed in Bayer's research laboratory, and he wondered how this figure compared to Kodak's research organization. Eastman was apparently too embarrassed to tell Duisberg that he had no research laboratory and employed only a few chemists in development (Sturchio 1985). At the same time, Eastman had grown alarmed about reports that Kodak was a subject of investigation by the Justice Department's beefed-up antitrust division. In April 1912, Louis Brandeis, then an antitrust advocate and outspoken critic of big business (later Supreme Court justice), addressed the City Club in Rochester, saying that large companies "wouldn't do any research because they were self-satisfied with their positions and didn't need any technical advice" (Sturchio 1985). Within a few months, Eastman had established a research laboratory and recruited its first director.

Whether Eastman took this step merely to bring his company into vogue (unquestionably, fads have occasionally characterized twentieth-century U.S. business practices), or whether he resolved to build a first-class research laboratory because he realized that the country's growing antitrust climate and threats of international competition made rationalization imperative can be debated.[26] Whatever his reason or motivation, the research laboratory at Kodak Park, directed by C. E. Kenneth Mees, the brilliant English chemist, provided the company with the basis for successful innovation for a long time. Under Mees's guidance, Kodak's R&D program played a major role in Eastman's move away from horizontal combination toward vertical integration. This shift served not only to move the company away from increasingly suspect business practices but also to lessen Kodak's dependence on German intermediates, photographic chemicals, and other products (Jenkins 1975, 322). Mees also wrote the first textbook on managing industrial research to be published in the United States, if not in the world (*The Organization of Industrial Scientific Research*, 1920).[27]

Several patterns, consistent with those elaborated by Alfred D. Chandler in *Scale and Scope: The Dynamics of Industrial Capitalism* (1990), emerge from this quick survey of R&D pioneers.[28] The leading firms in science-based industries created formal R&D programs at roughly the same time and for many of the same reasons. First, competition and the threat of having their core technologies undermined led executives, often in response to the urging of technically and scientifically gifted managers, to establish research and development programs independent of manufacturing and sales functions. These new R&D units were charged with providing for the long-term security of the firms' technology—industrial "life insurance," as Willis Whitney called it (Reich 1985, 37). Second, these pioneer industrial R&D programs were frequently created in the context of federal antitrust action; they were part of a movement to rationalize and integrate vertically companies that corporate leaders believed would overcome the objections to large-scale industry that had surfaced at the end of the nineteenth century. Third, these firms invested in R&D for the same reason that they invested in manufacturing and marketing; investment in R&D was part of the development of managerial hierarchies in large-scale American enterprise. Thus, the R&D function (the generation and application of knowledge) was internalized within the firm rather than left solely to the market. Finally, by contributing to what management theorists have termed "organizational capabilities," industrial R&D played an important role in the diversification and long-run success of these organizations in the twentieth century.[29]

Although the creation of R&D units may have signaled the decline of the myth of the independent inventor, it did little to lessen the effects of the "pure" science ideology on industrial research at any firm. Suspicion about the quality of science pursued in industrial research persisted. Indeed, throughout most of the twentieth century, research directors like Whitney have had to find ways to give industrial re-searchers the semblance of an academic research environment simply because industrial research is seen by academic elitists as a poorer career option than that offered by a university or a private basic research institute.[30] To attract scientists, corporate research directors have adopted liberal (though seldom unrestricted) publication policies and given their best researchers a great deal of latitude in choosing problems to work on. Even so, research directors have always had to keep their policies within the general bounds of company management practices and consistent with the technical needs of their firms.

The views of James B. Conant (1893–1978), a well-known organic chemist, head of Harvard University's department of chemistry, presi-dent of Harvard, and a major shaper of World War II research policy, suggest how leading academicians and many of their students ap-praised industrial research in the 1920s and 1930s.[31] On November 26, 1927, Conant's student, Louis Fieser, wrote to Conant about the DuPont Company's offer of an academic-style fundamental research position, saying, "I never expected to go into industrial work but the thing which makes a decision so difficult in this case is that I don't have to sell my soul at all; they even said that I could bring my quinones [a class of organic chemicals] along and continue my present work" (Conant Papers). Having turned down the DuPont offer in order to stay at Bryn Mawr College, Fieser wrote Conant on April 6, 1929, about the excellent work being done by DuPont chemists: "The industrialists [i.e., the DuPont chemists] are really a keen lot, though I don't think that they compare with us academics. . . . They impress me particularly as lack-ing in the fine critical judgment of the best teachers, and I wonder whether this is the cause or effect of their industrial relations" (Conant Papers). After giving a seminar at DuPont's Experimental Station, Conant himself was forced to note in a letter to Fieser, dated April 9, 1929, that "the crowd at du Pont's were the first people I met who seemed to have read my papers intelligently" (Conant Papers). Even so, Conant almost turned down a simple consulting arrangement with DuPont because, as indicated in a series of letters to Elmer K. Bolton dated June 6, June 18, July 12, and July 15, 1929, he worried about being tainted (Conant Papers). Letters dated June 1 and July 11, 1929, from Roger Adams assuring Conant that DuPont would treat him fairly and would not injure his reputation seem to have done little to ease his concerns (Conant Papers). Although Conant displayed great concern that his own reputation and the reputations of his best students would suffer from association with DuPont's industrial research program, he was not averse, he noted in a letter to Bolton on May 2, 1927, to recommending "a much less able man" for one of DuPont's positions (Conant Papers). Conant's actions indicate that industrial R&D programs were outlets for the growing numbers of scientists turned out by graduate programs at U.S. universities who were judged by their mentors not to be suitable for academic careers, hence giving industrial researchers a kind of second-class citizenship.

The Development of Private Research Institutes

Accompanying the rise of R&D units in some firms was the development of private, contract research laboratories. Their establishment reflects both the growing enthusiasm for industrial research in the United States and the maturing of the American research system. In 1886, Arthur D. Little, an important advocate of industrial research, established a firm that carried out chemical analyses for various companies that did not have these in-house analytical capabilities. Little's company then began to conduct larger studies for corporate clients of the strategic possibilities of new technologies and actually became involved in some development projects for clients. The company was not widely imitated at first, but it eventually met with competition from both private for-profit organizations and private nonprofit research institutes.[32]

The Mellon Institute in Pittsburgh, Pennsylvania, represented yet another development in the American system of industrial research. The institute came into existence when the Pittsburgh-based bankers and venture capitalists (and brothers) Andrew and Richard Mellon brought sometime-chemist, sometime-prophet Robert Kennedy Duncan to the University of Pittsburgh in 1910 with the idea of replicating and expanding upon what Duncan had done at the University of Kansas. There, Duncan had created and promoted (with less success than the Mellons were led to believe) a program of "industrial fellowships."[33] Firms would support advanced graduate students, who in turn would conduct project research for the firms as part of their graduate studies. With the generous financial support of its benefactors, the Mellon Institute was established in 1913 to do contract research and operate a program of industrial fellowships. Although these two sources of revenue were supposed to make the new institution self-sustaining, the institute always lost money, despite an impressive early growth of contracts and industrial fellowships. Even so, the Mellon Institute won high praise and attracted imitators. Most such private contract research firms have played an important role in the American system of industrial research.

As David Mowery has noted, although the Mellon Institute and its imitators were created in large part with the idea of providing R&D capabilities to smaller firms that could not establish their own laboratories, in practice, these institutes were used more by firms with strong in-house research capabilities (DuPont, Kodak, etc.) than by firms without internal research capabilities. Firms with their own R&D organizations usually employed private research institutes such as Mellon to supplement their own research, that is, to handle what they deemed routine types of research (Mowery and Rosenberg 1989, 79–82; Mowery 1983, 351–74).

By the time the Mellon Institute was established, Arthur Little had become president of the American Chemical Society (ACS). He had also become a major and vocal proponent of organized industrial research. Invariably he held up as a model of industrial research the German chemical industry, which then totally dominated the world markets in dyestuffs and pharmaceutical products. In a widely reprinted ACS presidential address in 1913, Little celebrated how U.S. industry had taken up industrial research and detailed instance after instance of American triumphs in industrial research, implicitly comparing them to those of Germany. Little estimated that there were, in addition to a small number of private institutes and trade associations, about fifty U.S. firms with formal R&D organizations (Little 1913, 793–801).[34] A close reading of Little's speech suggests American industrial research was then in its early adolescence. The needs and opportunities posed by World War I provided the context in which the American system of industrial research would rapidly mature.

THE MATURING OF INDUSTRIAL R&D

The outbreak of war in Europe in 1914 and the eventual entry of the United States into the conflict in 1917 were the catalysts for the rapid growth of industrial R&D in the United States. The British embargo of Germany first threatened and then choked off the supply of dyestuffs and pharmaceuticals to the United States, which was almost wholly dependent on Germany for these and other chemical products. Consequently, the U.S. government developed and implemented a program to promote domestic industries that could produce dyestuffs, pharmaceuticals, and other synthetic organic chemicals during the war and compete with the German firms after the war. Firms that already had research capabilities, such as DuPont and Kodak, used the war to diversify their businesses and acted as models for other firms that wished to establish R&D programs. The chemical industry got a further boost from the research conducted by and for the Chemical Warfare Service and from the industry's manufacture of several types of poison gas.

The pharmaceutical industry also benefited from some of the work done within the Chemical Warfare Service and from the need to produce drugs deemed vital for the nation's defense, such as the German-made, proprietary antisyphilis drug Salvarsan.[35] Although one or two American pharmaceutical companies had established some type of R&D program before the war (most notably Parke-Davis in 1902), the United States had only meager drug synthesis and development capabilities. The threat of a cutoff led to significant new programs in the U.S. pharmaceutical industry and in American pharmacology, many of which were initiated by the federal government. Like the nascent American dyestuffs industry, the pharmaceutical industry benefited from the confiscation of German drug patents by the Alien Property Custodian and their open licensing through the Chemical Foundation. (There is nothing like confiscating manufacturing plants, patents, and trademarks and then implementing a large tariff to help an industry get started.) After the war, most of the large drug firms built extensive R&D laboratories and began major programs of collaboration with universities, which soon led to some important drug developments.[36] Although pharmaceutical firms committed themselves heavily to research, the academics in the American Society for Pharmacology and Experimental Therapeutics, sharing a pure-science idealism with physicists like Henry Rowland, banned industrial scientists from being members of the society until the 1940s (Swann 1988; Parascandala 1990).

The chemical and pharmaceutical industries were not the only ones affected by the war. Because submarines, airplanes, and wireless communications were used with varying degrees of effectiveness, the country needed to develop these instruments of war and methods to counter them. The profound impact of these needs on developments in industrial R&D can be seen in several areas. First, as Daniel Kevles has noted, scientists in the country's research universities seized on the war in Europe and the perceived lack of American preparedness as the opportunity to advance scientific research in America. "I really believe this is the greatest chance we ever had to advance research in America," exclaimed George Ellery Hale, the astrophysicist and creator of Caltech, after the National Academy of Sciences (NAS) voted in 1916 to place the nation's scientific elite at the disposal of the federal government (Kevles, 1978, 112). With A. A. Noyes and Robert A. Millikan, two other outstanding scientists whom he would later attract to Caltech, Hale hoped to transform NAS into the central body for scientific research in the United States. Research, Hale hoped, would be conducted and directed by NAS on a scale and at a level of quality that would rival that of any nation in the world. Because they deemed NAS moribund, Hale and his cohort maneuvered to create the National Research Council (NRC) as the research arm of NAS but under the cohort's control. Their success in 1916 stemmed from their theory—and its acceptance by the Wilson administration—that science and scientists could contribute to the war effort. To help their cause (indeed, to legitimate it) Hale, Millikan, and Noyes enlisted the support of GE's Whitney, who had been Hale's classmate and Noyes's student and later colleague at MIT (Kevles, 1978, 109–11).[37]

Throughout the entire period of war preparations and much of the war, the NRC's principal goal was to build the kind of scientific establishments that Germany already had in its universities and other research institutions. This was a case of pure opportunism on the part of Hale, Millikan, Noyes, and their followers. Indeed, Whitney's biographer George Wise stresses that

> during that critical year of 1916, Whitney and Hale spent as much or more time discussing how to get the government permanently committed to supporting science than they spent considering how to find submarines or fix nitrogen. . . . [Their] efforts [to promote research] distracted American scientists from the job of mobilizing science for defense in 1916. . . . Whitney and his contemporaries put the job of promoting science first with nearly fatal consequences to the defense technology effort. (Wise 1985b, 186–87)

This is not to say that Hale, Millikan, Noyes, and other university scientists who temporarily left their academic posts to serve the nation did not succeed in bringing science to bear on such difficult problems as chemical warfare and submarine detection. Certainly they did. Their success convinced them that professional scientists could contribute more to weapons development than could such organizations as the Naval Consulting Board (NCB), which Secretary of the Navy Josephus Daniels created in collaboration with Thomas Edison in 1915. The NCB was made up of leading inventors and representatives from most of the professional engineering societies, but it lacked representation from the NAS and some of the nation's other leading scientific societies, such as the American Physical Society. The NCB screened ideas for weapons submitted by American citizens in response to Daniels's and Edison's appeal for the ideas of the nation's best and brightest inventors.[38] Americans submitted some 100,000 ideas for weapons, but the NCB found only one or two worthy of further consideration. The age of the heroic inventor, if that age had ever existed, was finished.

While the NCB may not have contributed to weapons development, it did have one long-lasting impact on American research. Backed by the other members of the board, Edison convinced Daniels and Congress to create a research laboratory within the navy. Today's Naval Research Laboratory traces its origins to World War I and the NCB. The construction and staffing of the laboratory shortly after the war was something of a miracle considering that many of the schemes and plans that had been hatched during the war were abandoned after the armistice. The Naval Research Laboratory carried out important research in the interwar period, including some pioneering work on radar (Allison 1981).

Another World War I period creation survived the war, and it had an important impact on the development of both civil and military aviation. This was the research program and facilities of the National Advisory Committee on Aeronautics (NACA). At its research facilities, which included large wind tunnels and other equipment, NACA generated an enormous amount of fundamental knowledge in aerodynamics, airframe design, aircraft testing, and many other areas. NACA served its patrons and clients so well that it was transformed into the National Aeronautics and Space Administration (NASA) when the space race began in the late 1950s.[39]

Within American industry, the war also brought widespread attention to the existence, organization, management, and achievement of the research laboratories of the R&D pioneers. Industrial research programs came under the purview of the NRC, which established the Advisory Committee on Industrial Research. The committee initiated the nation's first efforts to compile statistics and other information on industrial R&D programs. The founding research directors at GE, DuPont, Kodak, and AT&T—Willis Whitney, Charles Reese, Kenneth Mees, and J. J. Carty and Frank Jewett—moved into the limelight as they advised military, government, and industry figures on how scientific research could be harnessed for technological development. These men became national leaders in the field of industrial R&D.[40] Working under the NRC's banner to promote industrial R&D, Reese, Whitney, and others published pieces on the benefits and management of industrial research. Kodak's Mees, for example, published an influential article in *Science* that became the basis of his textbook on the management of industrial R&D. In the article, Mees stressed that firms and industries "must earnestly devote time and money to the investigation of the fundamental theory underlying the subject in which they are interested" (Mees 1916, 766). These efforts led several firms to establish research programs modeled on the programs of Kodak, GE, and other R&D pioneers. Westinghouse Electric, for example, founded a basic research program as prescribed by Mees and went so far as to recruit a Kodak scientist to manage the program.[41]

Whitney and his counterparts at AT&T brought together their firms' best researchers at government facilities at Nahant, Massachusetts, and New London, Connecticut, to work on the problem of submarine detection. The creation of two separate facilities reflected a dispute between the NCB and the NRC—a dispute that left Whitney caught in the middle because he was a member of both organizations. Despite turf battles, however, industrial researchers and academics carried on. Whitney assigned Irving Langmuir, who had already done the work for which he later won the Nobel Prize, and Charles Eveleth, who would later take charge of all of GE's manufacturing, to work on the problem. Two future presidents of Bell Laboratories, Harold D. Arnold and Oliver E. Buckley, were part of the AT&T contingent organized by Jewett. Millikan himself, who had said that submarine detection was a "problem of physics pure and simple," worked with these industrial researchers and learned firsthand about the strengths and limitations of project research undertaken by top-flight scientists and electrical engineers. As head of the NRC's subcommittee on submarine detection, Millikan monitored the work at the New London Experimental Station, which at the time of the armistice in November 1918 was staffed by 32 university professors and 700 enlisted men. The teams at the two installations did succeed in developing submarine detection equipment, which helped to solve one of the Allies' major problems and eased public fears of the German U-boat menace (Kevles 1978, 116–26; Wise 1985b, 187–94).

After the war Whitney, Reese, Mees, and others formed the Directors of Industrial Research (DIR), which is still very much alive today. Not affiliated with NAS, NRC, or any government agency, the DIR provided a critical forum in which research directors could informally share information on such topics as starting salaries for scientists with doctoral degrees, publication policies, and coordination of research in a diversified firm. In addition, members of the DIR hosted one- and two-day tours of their firms' research facilities. These sometimes elaborate events not only served as a means of seeing what kind of R&D facilities other corporations had built but also gave less-fortunate research managers ammunition to use in convincing their firms' executives to expand research facilities and programs (Hounshell and Smith 1988, 315–16).

Without question, then, World War I led to a widespread quickening of interest in and enthusiasm for industrial R&D in the United States. At the same time, the war gave the inheritors of Henry Rowland's "pure" science idealism the wedge with which they hoped to open up institutional, governmental, industrial, and public support for "pure" science.

The Interwar Period

Although most Americans sought what President Warren G. Harding called a "return to normalcy" after the war, many academic scientists did not. The research infrastructure that Hale, Millikan, and others had started to put in place was by no means complete and was threatened by "normalcy." These scientists pledged themselves to exploiting existing channels for achieving "best science" as well as to opening up new ones. They won a permanent charter for the NRC as a government-sanctioned (but not financed or controlled) body acting on behalf of American science. They also found support from the Rockefeller Foundation, which donated $500,000 for doctoral and postdoctoral research fellowships governed by the NRC (Kevles 1978, 155–56).

Industry also began to do its part vis-à-vis science education. DuPont was among the first companies to offer "no strings attached" fellowships. In 1918, at the urging of Charles Reese, DuPont's research director, the company gave 18 fellowships and 33 scholarships to institutions with strong programs in chemistry and chemical engineering. By 1940, some 200 companies were offering more than 700 fellowships. These fellowships were fundamentally different from the fellowships offered by the Mellon Institute.

Hale, Millikan, and others then moved to create a German-style research institute, a national scientific laboratory devoted to basic research and managed and controlled by scientists. In 1919, the Carnegie Corporation gave Hale $5 million, one-third of which was to be used to build a new home for the NAS and the remainder of which was to be used for NRC activities. Hale tried to supplement the Carnegie endowment with federal monies for NRC research programs, but it soon became clear that neither Congress nor the administration wanted to support basic science without political and programmatic controls. Although the NRC positioned itself to address issues deemed vital to the nation, it did not function as Hale had envisaged.

Having failed in Washington, Hale, Millikan, and Noyes shifted their focus to a sleepy technical institute in Pasadena, California, and transformed it into a major scientific and technological research university: the California Institute of Technology, Cal Tech for short (Kevles 1978, 155–56; Kargon 1982, 100–50; Servos 1990, 263–66; and Goodstein 1991, 64–108). The cohort's success in building up Cal Tech stemmed from selling "best science" as the only solution to the emerging global competition in industry in the postwar period. Hale's warning in 1916 that after the shooting stopped Germany would conduct an "industrial war" with the Allies resonated with Millikan's postwar analysis that "the War has taught the manufacturer that he can not hope to keep in the lead of his industry save through the brains of a research group, which alone can keep him in the forefront of progress."[42] In a book called *The New World of Science,* which was edited by Robert Yerkes (1920), Hale and Millikan exploited the gains made during the war to seek substantial private and foundation support for science and scientific research at Cal Tech. By pursuing "best science" at Cal Tech, they and their colleagues would serve the nation in peace as they had in war. Hale and Millikan succeeded with their objectives at Cal Tech, and other scientists prospered at their own research universities. But the advocates of "best science" fell short in winning permanent, unfettered, unquestioning support for "pure" science, from either the government or industry (Tobey 1971; Kevles 1978; and Kargon 1982).

This failure is best illustrated by the effort to build a national research endowment in the mid-1920s. Herbert Hoover, who was then secretary of commerce and a hero for his coordination of war relief in Europe, led a campaign, conceived by Hale, Millikan, and other NRC members, to raise an endowment of $20 million, principally from industry. Hoover became a champion of "pure" science, arguing that both industrial and social progress depended on it (Hoover 1926, 6–8). However, despite appeals by Hoover and other members of the fund's board of trustees, which included Andrew W. Mellon, J. J. Carty, Owen D. Young (an "industrial statesman" and major supporter of GERL's Willis R. Whitney), and Julius Rosenwald (the chief executive at Sears who had supported the creation of product testing laboratories within his company), among others, there were few corporate donors. The effort would be declared stillborn in 1932.[43]

Even so, Hoover's message about "pure" science served as a catalyst, pushing several corporate R&D programs either to build fundamental research into their ongoing R&D programs or to increase spending for "pure" research. Indeed, statements by such research leaders as Whitney and Mees, who declared that enormous dividends were available to any industrial company that would found a scientific research laboratory, and a national campaign for efficiency engineered by Herbert Hoover as secretary of commerce created a national frenzy for industrial R&D. By 1925, the GERL was known nationwide as the "House of Magic" in which scientists carried out modern miracles (Wise 1985b, 214). Further stimulating the frenzy was the NRC's campaign, run by the director of its division of engineering, Maurice Holland, to sell industrial R&D as the "royal road to riches" (Wise 1985b, 215; Kevles 1978, 170–84). Holland and the NRC would go on to establish in 1938 the Industrial Research Institute (IRI), an association of industrial research programs and managers, which was incorporated as an organization independent of the NRC in 1945.[44] The IRI continues to engage in both research on industrial research and lobbying on behalf of its members.

Between 1919 and 1936, some 1,150 industrial research laboratories were established by U.S. manufacturing firms. This is roughly 54 percent of all industrial R&D laboratories founded from the beginning of industrial research until 1946. In 1921, 2,775 research professionals (scientists and research engineers) worked in industrial research establishments. This figure grew to 6,320 in 1927, 10,927 in 1933, and 27,777 in 1940 (Mowery and Rosenberg 1989, 61–74; Weart 1979, 306). Even in the Great Depression, therefore, industrial research grew rapidly.

Given the campaigns to promote industrial research and the widespread public attention these efforts generated during the 1920s and 1930s, leaders of technologically-based corporations that did not have formally designated R&D organizations felt out of step. Such was the case at United States Steel. From the time of its formation in 1901 as a result of the largest, most famous, and most expensive merger in U.S. history, U.S. Steel had carried out all its developmental work in technical organizations located throughout the company's various divisions, which corresponded closely to the separate companies that had merged. The divisions did virtually no scientific research, and no research was done at the corporate level.[45] Unlike DuPont and dozens of other companies that had grown or been created through mergers, U.S. Steel had not consolidated its operations by closing down inefficient plants; it did not coordinate manufacturing across the formerly independent companies until long after the merger; and it had not built R&D organizations as part of a process of rationalization. When U.S. Steel created its research organization between 1926 and 1927, it did so because industrial R&D had become fashionable.[46]

U.S. Steel's managers and directors knew almost nothing about scientific research either inside or outside the universities or about industrial R&D in general.* This was true even though there were a large number of product development and control laboratories spread throughout the company's many divisions and plants. In the mid-1920s, the board of directors decided that, given all the attention Hoover and others had called to research in industry, U.S. Steel ought to have a central R&D organization that would do scientific research. They appointed George Crawford, the president of U.S. Steel's Tennessee Coal, Iron, and Railroad Company, to search for a suitable candidate to head the laboratory. Seeking advice, Crawford visited Frank B. Jewett, president of Bell Labs, the largest industrial R&D organization in the world. Coincidentally, Robert Millikan was also visiting Jewett. Together, the two men recommended four possible candidates. Millikan also agreed to advise Crawford and the steel company on how to set up a research laboratory.

Crawford interviewed the four people recommended by Jewett and Millikan as well as seventeen other scientists. In March 1927, after receiving a four-page report from Millikan about how to organize the projected lab, Crawford and his committee unanimously recommended Millikan himself as the new research director. At this time, Millikan was president of Cal Tech, had recently received the Nobel Prize in physics, and was a major proponent of "best science" in the United States. Minutes of the board meeting make it clear that the members knew little about Millikan. Some were more concerned about his ability to play golf than his Nobel Prize. Judge Elbert Gary, chairman of the board, sent a telegraph to Millikan asking him to come east to talk to the board without telling him why. Incredible as it sounds, these steel men assumed that, once they had assured themselves that he was an

---

* This section on U.S. Steel is drawn from Paul A. Tiffany, "Corporate Culture and Corporate Change: The Origins of Industrial Research at the United States Steel Corporation, 1901–1929," paper presented at the Annual Meeting of the Society for the History of Technology, Pittsburgh, 1986.

"all-right fellow," Millikan would accept their invitation to become the founding director of U.S. Steel research. They understood neither Millikan nor the politics of science in the United States. A gentleman and a diplomat, Millikan refrained from laughing and politely refused. The board, again showing its profound ignorance of research and swollen sense of U.S. Steel's importance, then offered the job to Jewett, who, of course, also declined.

Both Millikan and Jewett again offered the board their help in recruiting a good scientist, and shortly thereafter, at Millikan's and Jewett's urging, Yale University chemist John Johnston, one of the four they had originally recommended, signed on with U.S. Steel. Poor Johnston was not aware until later how little the company's board knew or cared about research and how little support he would receive. His research division was consigned to an old industrial building in Kearney, New Jersey, and it received far less research money than Johnston had initially been promised.[47]

How many of the hundreds of firms that established R&D programs in the 1920s did so for the same reasons as U.S. Steel? How many did so because they saw real short- or long-term benefits from industrial R&D? In the absence of reliable histories of individual cases, the answers to these questions are uncertain, but as Mowery's figures make clear, the number of R&D programs and researchers in those programs rose markedly during this period and, despite some large layoffs in the early years of the depression, showed major growth in the 1930s as well (Mowery and Rosenberg 1989).

The interwar period is also noteworthy for the significant science that came out of some of the industrial R&D programs. Of course, GE's Irving Langmuir had already done the research that led to his Nobel Prize in 1932, but he continued to turn out excellent science, which helped GE recruit other outstanding scientists (Wise 1985b, 149–66). In 1927, only two years after AT&T had formally incorporated Bell Labs, Clinton J. Davisson began the work on electron diffraction that led to his 1937 Nobel Prize in physics, the first of many Nobel Prizes won by Bell Labs researchers. Davisson's science neither led to revolutionary new products nor did it dramatically transform Bell's technology, but it furthered Bell Labs' understanding of electron behavior by confirming that electrons were diffracted by a target (such as an element in a vacuum tube) in the way predicted by wave mechanics. Moreover, Davisson and Germer's publication on electron diffraction in the *Physical Review* in 1927 served as a clear sign of the quality of science done at Bell Labs (Russo 1981).[48]

Occasionally outstanding fundamental research not only produced revolutionary new products but also fundamentally altered a company's research program. In 1928, a young organic chemist named Wallace H. Carothers left an instructorship at Harvard to join a DuPont program devoted exclusively to fundamental research.* The company had recently established the program within its central research department, in part as a response to Hoover's call for more basic research. Carothers had been hired to head a group in organic chemistry, and he had been encouraged to focus his research on polymers. Although he had never worked in this area, Carothers was well aware that chemists were hotly debating the nature of polymers. One school of thought believed that polymers were aggregates of colloidal particles held together by special forces that did not operate like ordinary covalent chemical bonds. Another school, led by Hermann Staudinger, a German chemist, believed that polymers were "macromolecules" made up of ordinary molecules bound together in chains by ordinary bonds, but nobody had been able to prove this. In a little over two years, Carothers and his team of organic chemists managed to build a massively documented case in favor of the Staudinger view. Carothers published two landmark papers—"An Introduction to the General Theory of Condensation Polymers" (1929) and "Polymerization" (1931) —that developed the general theory of polymers, or macromolecules, and described and gave lasting names to the basic processes of polymerization.

Carothers's work was disseminated rapidly owing to the unrestricted publications policy of DuPont's fundamental research program at that time. Both the work and the freedom to publish brought critical acclaim to DuPont's research community, but in 1930, things changed. New leadership in the central research division, increased pressure stemming from the nation's worsening economic situation, and two very important discoveries by Carothers's group hastened the change. The first discovery was a polymer, which Carothers named chloroprene, that had a chemical formula directly analogous to that of isoprene (natural rubber). Chloroprene possessed rubberlike properties and could be processed in the same way as rubber. More importantly, unlike rubber, it was highly resistant to degradation by gasoline and oil, which suggested that it had important marketable properties. Carothers and his group had synthesized the world's first wholly synthetic rubber, later known as neoprene. In that same month (March

* This section on Carothers's and DuPont's fundamental research is based on David A. Hounshell and John Kenly Smith, Jr., *Science and Corporate Strategy: DuPont R&D, 1902–1980* (New York: Cambridge University Press, 1989), 223–74.

1930), owing to some creative laboratory techniques, Carothers's group polymerized a high molecular weight aliphatic polyester and discovered that it could be spun into a filament and then "cold drawn" (stretched, which aligned the polymeric chains, thus yielding much finer and stronger filaments). This was the world's first wholly synthetic fiber.

These two discoveries transformed Carothers's research program from one devoted to "pure" research to one focused on "pioneering" research. Instead of seeking deeper knowledge about polymers or polymerization, Carothers and his group were now dedicated to finding commercially viable products. They published twenty-three papers on chloroprene, which Carothers noted in 1932 were "abundant in quantity but a little disappointing in quality" (Carothers 1932). They also worked to synthesize a polymer that could be spun and drawn into a fiber having properties qualitatively similar to natural fibers but possessing a high melting point and stability in water (unlike the first group of synthetic fibers they had polymerized and spun in March 1930).

When the work on neoprene was turned over to DuPont's organic chemicals department, which had its own significant research capabilities in organic synthesis and rubber chemistry, the pursuit of a synthetic fiber remained with Carothers's lab. After working to find a polymer with all the right properties and failing, Carothers returned to more fundamental research and did some interesting work on large-ring compounds (as opposed to simple linear polymers), which he published in the leading chemical journals. However, when this research was finished and Carothers seemed uncertain where to go next, his research director—who had once been opposed to DuPont's undertaking any fundamental research—encouraged him to return to the synthetic fiber problem, which he did. In a matter of a few weeks, Carothers's team, pursuing a synthesis route laid out by Carothers, discovered a class of polymers known as polyamids; the product that emerged from this work—nylon—gave DuPont its first commercial blockbuster. Put on the market in 1939, nylon, by the 1990s, had earned the company as much as $20 to $25 billion.

By 1939, then, a formula for industrial research laboratories seemed to be emerging, at least at DuPont and soon in other companies. The formula seemed quite simple: Do world-class fundamental scientific research, and you will find important new products that you can then commercialize and profit from enormously because they are completely proprietary.[49] World War II, which had already begun in Europe by the time DuPont's commercial nylon plant started operations, served to confirm and ratify this formula a dozen times over. The war years would be a decisive watershed in the history of industrial R&D in the United States.

World War II and the Postwar Period

World War II produced the Age of Big Science. The atom bomb, radar, the proximity fuse, antibiotics, the digital electronic computer, and numerous new materials, theories, and analytical techniques all emerged from the concentrated efforts of university scientists brought together to address wartime needs. The scientific community in the United States had been looking for such an opportunity since the end of World War I. Scientists who had been trying to counter arguments that scientific advance led to massive unemployment during the Great Depression (Weiner 1970, 31–38; Kevles 1978, 236–66), came out of the war as heroes, not as revered as Eisenhower or MacArthur but nonetheless held in awe.[50]

The shibboleths of this new age were that basic science and well-funded scientists produced dramatic new technologies and that scientists knew better than generals, engineers, or industrialists what new science to pursue, which new technologies to develop, and how best to deploy those new technologies.[51] Seldom have the lessons of war been more fundamentally misunderstood. Seldom have such misunderstandings been more important, for they governed the course of national policy and the direction of U.S. industrial R&D until the 1960s.

What everyone, including those who should have known better, overlooked was that none of these new technologies and products could have emerged without the enormous engineering and manufacturing know-how and capabilities of the nation's corporations. Even the Manhattan Project depended on the advanced engineering, construction, and operational capabilities of numerous firms, from the engineering firm of Stone & Webster to manufacturing companies such as General Electric, Westinghouse, Union Carbide, M. W. Kellogg, and DuPont.[52] Even more overlooked was the degree to which the nation's capabilities in mass production, rather than the actions of a group of brilliant physicists on a mountaintop in New Mexico or in a laboratory in Chicago, determined the course of the war.[53] In numerous newspaper and magazine articles and in such uncritical books as James P. Baxter's Pulitzer Prize-winning *Scientists Against Time* (1946), scientists were portrayed as the saviors of the nation.[54] This misreading was further compounded by the carefully orchestrated actions of the nation's scientific elite, who sought to achieve the permanent support for science that they had been trying to garner since the middle of the nineteenth century.

Paradoxically, an electrical engineer, Vannevar Bush, who prized fundamental research, took the leading role in U.S. science policy during and immediately after the war. With the assistance of James B. Conant, Bush had directed the nation's wartime efforts in science and technology. Seeking to head off Senator Harley Kilgore and the Subcommittee on War Mobilization's postwar science policy, which threatened to democratize science and to address the problem of concentration of federal arms contractors, Bush manipulated the Roosevelt White House into requesting a study on science policy (Kevles 1978, 341–48; Kevles 1977, 5–26; and Kevles 1990, ix–xxxiii). He then assembled a team of scientists who shared a "pure" science ideology. Together, he and his team issued the document that established the new catechism. Bush played on the beliefs of two generations of Americans who had been taught that the frontier had shaped the country's destiny, character, and institutions and titled the report *Science—the Endless Frontier* (1945).[55] The report promised Americans an endless frontier through science. By building and supporting institutions committed to "best science" and governed by scientists, Bush and his team wrote, Americans would reap an endless bounty of new technologies that would ease their burdens and bring wealth to the country and stability to the world. Without these institutions, they went on to say, Americans faced dim prospects because the war had seriously harmed many of the European scientific institutions on which American science and industry had depended.

A number of historians have described and analyzed how Bush's agenda was played out in Washington after the report was issued in July 1945 (Kevles 1978; Sapolsky 1990; and Geiger 1992). Their work suggests that the creation of the National Science Foundation (NSF) in 1950 was a pyrrhic victory for Bush. He had promoted the creation of a national research foundation to support basic research in universities, but he had then done battle for five years with President Harry S Truman, who sought democratic control of government-sponsored science. As Kevles (1978, 353–64) and others have shown, by 1950 the military, once critical of "wild-haired scientists," had become the largest funder of basic research in the United States.[56] Indeed, after 1946 the U.S. Navy became *the* major government funder of basic science in the country. Bush had run roughshod over the navy during the war, pushing it out of participation in the Manhattan Project and keeping other critical projects out of its purview. In response, the navy created the Office of Naval Research (ONR) at the end of the war and began a program of funding basic research in universities, which its leaders thought critical to the long-term development of a nuclear navy. Although university scientists who had worked on military projects during the war were at first fearful of the red tape, compartmentalization, and strictures against publication that had characterized wartime research, they soon discovered that the navy was offering them big research contracts largely free of these impediments. Moreover, much of the research funded by ONR seemed so far from any military application that it truly could be seen as "pure" (Kevles 1978, 353–96; Sapolsky 1990, 37–56).[57] Soon, the army and the newly created air force were following the ONR's lead, offering civilians research contracts for research while establishing their own in-house research and development departments. (Komons 1966; Sturm 1967; Sigethy 1980; and Gorn 1988).

While Bush and Truman and their followers may have battled over democratic controls of science, no one ever questioned Bush's linear model of new science as the principal source of new technology. In fact, with the exception of Bush's belief that all basic research should be done in universities, American industry firmly embraced the Bush model. In the postwar period, increasing numbers of companies that did not have R&D programs created them. In 1945, for example, Thomas J. Watson, Sr., the founder and president of International Business Machines (IBM), created a department of pure science within his company and hired the astronomer Wallace Eckert to direct it (Pugh 1986, 522–70). In September 1949, Ernest Breech, a former manager of General Motors' Bendix Division who had been hired to help rebuild the ailing Ford Motor Company, wrote in a memorandum to Henry Ford II, "I am convinced that Ford will not have many 'firsts' unless we get a few good thinkers and have a real research department" (Breech 1949). By "real research," Breech clearly meant scientific research. In mid-1951, Ford established within its Dearborn Engineering Laboratory a new "scientific laboratory" that was dedicated, according to its founding director, to "fundamental research and development in fields broadly related to the basic character of Ford Motor Company—transportation" (Ford Industrial Archives). Soon basic research in physics and chemistry was being pursued at Ford, a marked departure for a company that had been built around the homespun practicality of the first Henry Ford.[58]

Firms that already had R&D programs dramatically expanded them and directed their expansions at fundamental research.[59] DuPont provides a good example of the new paradigm's power. Even before Pearl Harbor, DuPont's executives had begun to plan for the postwar era.* Their thinking was conditioned by at least two developments: the dramatic expansion of the Antitrust Division of the Department of Justice and the continued anti-big business sentiment in critical circles of Congress and the Roosevelt administration; and the tremendous success of nylon. The validity of the research formula derived from the company's experience with nylon after 1939 grew more certain when the company was selected by General Leslie Groves to design, build, and operate the Hanford Engineer Works, which produced plutonium for the Manhattan Project, and executives learned more and more about wartime science projects.[60] After the Department of Justice brought suit against DuPont and Britain's Imperial Chemical Industries for violation of the Sherman Act in 1944 and then charged the company with several other antitrust violations, executives arrived at the company's postwar strategy for growth: increase the company's basic research and watch blockbuster products emerge. In discussions and memoranda written between July 1945 and early 1946, company executives employed language that suggests that they had also carefully read Bush's report and accepted its call for more science.

In October 1945, Elmer K. Bolton, the director of DuPont's central research department who commanded enormous respect throughout the company and in the DIR, reported to the president of the company that "the country is about to enter a period of unparalleled scientific activity." He went on to note that research would be expanded in the universities and that industry was poised to "undertake a very substantial expansion of personnel and of research facilities" (indeed, a DuPont study had indicated that more than fifty new industrial R&D laboratories had recently been built, and far more were being planned). Moreover, the government's actions, through the army, navy, and Bush's proposed national research foundation meant that research would be conducted "on an enormous scale" (Bolton 1945). There would be no return to the "normalcy" that had followed World War I, Bolton wrote, because most people believed that science was the endless frontier and that all new technology was derived from basic research. To "retain its leadership," he argued, DuPont had "to undertake on a much broader scale fundamental research in order to provide more knowledge to serve as a basis for applied research" (Bolton 1945).

---

* The following four paragraphs are based on David A. Hounshell and John Kenly Smith, Jr., *Science and Corporate Strategy: DuPont R&D, 1902–1980* (New York: Cambridge University Press, 1988), 331–65.

At the time, DuPont was a decentralized, multidivisional firm. It had eleven operating (manufacturing) departments, each working in its own segment of the chemical industry under relatively autonomous leadership. Each department possessed its own manufacturing, marketing, and R&D divisions. In addition, there was the central research department (the one that had discovered and done much of the development on nylon). The company's central engineering department also had its own fundamental research program in chemical engineering. Over the objection of the vice president for research, Charles M. A. Stine, who had created central research's fundamental research program in 1927, DuPont's executives decided to expand each of the operating department's research programs and to commit a sizable portion of the expansions to fundamental research. Each department, it was said, would generate a stream of "new nylons" in its respective business (Minutes of the Executive Committee, E. I. du Pont de Nemours & Co., June 13, 1945). Arguing against this strategy, Stine noted that the R&D units of the operating departments had their hands full with their existing businesses and would not be able to run effective programs of long-term fundamental research. Moreover, he maintained that because many of the departments manufactured products that possessed a common scientific basis (for example, paints, fibers, plastics, and films were all polymeric in nature), departmental research probably would be duplicated in other research units as well as in the central research unit. He argued that DuPont should expand its fundamental research but asserted that it should be done only by the central research unit.[61]

Under the leadership of Crawford Greenewalt, DuPont's new chief executive officer who had also been one of the critical leaders in the nylon development project and had then served as the liaison between DuPont and the Metallurgical Laboratory at the University of Chicago, the company proceeded to expand its research dramatically, especially its fundamental research. The new basic research programs in all departments had a profound, if unexpected, impact on the company's central research unit. To avoid duplicating the work now being done by the operating departments, the central research unit undertook work that was far out on the horizon and in areas in which the company had little or no technical capabilities or business interests. Two decades of such programs would eventually make the central research unit vulnerable to the charge that its research, although impressive by academic standards, was irrelevant to the company. Indeed, some would charge that central research had become an ivory tower (Hounshell and Smith 1988, 574–85).

The DuPont experience was by no means unique. An examination of the histories of the other major R&D pioneers—General Electric, Kodak, and AT&T—reveals that virtually all of them committed themselves to the same type of research program expansion and to pledging more resources to fundamental research. They did so for the same reasons that DuPont did. They believed in the linear model and thought that it offered companies one legally defensible way to grow.[62] The wartime science projects, especially the Manhattan Project, and the widespread knowledge of nylon's origins and incredible success were too big and too real to be gainsaid by any company. After Charles Kettering, General Motors' "professional amateur" research and engineering director, retired in 1947, the company reoriented its research programs toward more basic research. By 1956, the company was ready to dedicate a building to its research programs. In May of that year, the new General Motors Technical Center was dedicated (Leslie 1983, 317–18). RCA also shifted toward fundamental research. The company expanded the RCA Laboratories that had been built in 1941 near Princeton University and, in the decade after 1945, hired more theoretical scientists. By 1955, scientists represented 50 percent of the lab's staff (Graham 1986, 68–71). Similar commitments to the linear model were made in the pharmaceutical industry. Merck & Co., which already possessed extensive research capabilities, expanded its basic research and put Vannevar Bush on its board of directors in 1949 (Merck & Co. Inc. 1991). Ever the champion of basic research, Bush later became chairman of the board.

Faith in the linear model and in its power to yield blockbuster products grew even stronger in the 1950s after the commercial importance of the transistor became apparent to a wider audience than Bell Labs' personnel. Histories of the transistor's discovery, written in an age of faith in the linear model, claimed the device was a product of pure research in an enlightened laboratory.[63] In fact, Bell Labs' reputation and the basic research of the other R&D pioneers led IBM to expand its research enormously. In 1956, IBM established a research division devoted to building a world-class basic research program, modeled in part after the program Westinghouse had established in the late 1930s. By 1960, the company had opened its Thomas J. Watson Research Center, a massive laboratory designed by the Finnish-American modernist architect Eero Saarinen (Pugh 1986, 522–70).[64]

The Cold War and Research

War's end did not spell the end of the special weapons research laboratories that had been established during the war. Initially these laboratories experienced peacetime declines in funding and personnel in the immediate postwar period. General Leslie Groves became alarmed by the exodus of scientists from the Manhattan Project laboratories and concluded that the federal government would have to initiate new relationships with university scientists to further the development of nuclear weapons (Geiger 1992). Yet, any idea of scientific demobilization disappeared when the Soviet Union began to give the United States problems on the eastern frontier. These problems led to the Truman Doctrine in 1947 and the confrontation over Berlin in 1948. The detonation of an atomic bomb by the Soviet Union in 1949 and the outbreak of the Korean War in 1950 brought about a scientifically and technically oriented arms race that rivaled the R&D projects of World War II. The facilities that had been created as temporary measures in World War II became national laboratories in 1946 and 1947, and entirely new laboratories were created for weapons development and production. Work on the hydrogen bomb was done at the rapidly expanding Los Alamos complex in New Mexico, and in 1952, the Livermore Laboratory was established near San Francisco as the nation's second nuclear weapons laboratory. As Robert Seidel has argued, even those national laboratories with missions to pursue "pure" science, such as Brookhaven National Laboratory on Long Island, New York, were authorized and sustained because of the overriding concerns about national security brought about by the cold war (Seidel 1983, 375–400; 1986, 135–75; 1990, 420–41; 1994, 361–91). In fact, in conjunction with the linear model, which it outlasted, the cold war played a fundamental role in shaping the course of industrial R&D in the postwar era. Only after it had ended were scholars and policymakers in a position to understand fully how deeply it had shaped the country's industrial R&D.

Early in the cold war, new generations of nuclear materials reactors were developed and built by General Electric, which had taken over the operation of the Hanford Engineer Works from DuPont in 1946, and by DuPont, which reluctantly agreed in 1950 to assume responsibility for a new plant in South Carolina that was intended to produce nuclear materials for hydrogen bombs.[65] Under the entrepreneurship and management of Admiral Hyman Rickover, the navy developed nuclear submarines and aircraft carriers, working closely with West-inghouse and General Electric (Hewlitt and Duncan 1974). The air force launched a program to develop an intercontinental ballistic missile system that could deliver atomic warheads to the Soviet Union. In the process, it created a new "systems-driven," high-technology firm named TRW, led by Dean Woolridge, a former Bell Labs researcher, and Simon Ramo, a former GERL member (Beard 1976; Dyer 1993).[66] Long-range, high-altitude bombers were developed by several aircraft companies that dramatically expanded their R&D capabilities and fa-cilities, largely at government expense. A computer-driven early warn-ing system was developed by IBM to protect the United States from a surprise attack by Soviet bombers coming over the North Pole. The system was based on the wartime and immediate postwar research of the Servomechanisms Laboratory at MIT (Redmond and Smith 1980). Other new countermeasures were also developed. Furthermore, new chemical and biological weapons emerged from many laboratories. The list of R&D projects for the military is seemingly endless, and to it must be added the manned space program of the National Aeronautics and Space Administration (NASA). As several historians have noted, the creation of NASA was driven more by the exigencies of the cold war than by any desire to explore the universe, and it led to the creation and expansion of numerous industrial firms acting as prime contractors and subcontractors to NASA (McDougall 1985; Levine 1982; and Koppes 1982).[67]

Without question, the cold war drove federal spending for research almost entirely in the direction of the military. As Kevles (1978) has noted, "For almost a quarter-century after 1945, defense research ex-penditures rose virtually exponentially, even in constant dollars, ac-counting through 1960 for 80 percent or more of the entire federal R&D budget. In 1950, it was estimated that there were 15,000 defense re-search projects; in the early 1960s, perhaps 80,000" (p. 466). Paul For-man (1987) has gone even further, writing that "through the 1950s, the only significant sources of funds for academic physical research in the U.S. were from the Department of Defense and an Atomic Energy Commission whose mission was *de facto* predominantly military" (p. 194). Forman has argued that the entire development of quantum electronics in the postwar period can be attributed to military funding of research.

Industry, of course, was also affected by this development.[68] As Forman has noted, the electronics industry in particular came to de-pend on the military for its R&D funding. In 1960, for instance, the federal government (almost exclusively the military) paid for 70 percent of the R&D conducted by the electronics industry (Forman 1987, 164–65). With the military funding much of its research, the electronics industry became increasingly conservative about the way in which it spent its own money in research. Most historians and electronics industry analysts now agree that the United States ceased to be the leader in consumer electronics in part because of its preoccupation with military electronics, which led companies to focus attention on performance objectives rather than market objectives.[69]

The growth in military spending for R&D, the overwhelming role of military funding, and the nature of that funding in advanced electronics, aeronautics and space, nuclear weapons, and related areas meant that companies found the costs of securing researchers mounting rapidly after 1947. This problem grew even worse with the development of the U.S. space program in response to the Soviet launch of *Sputnik* in 1957 and the panic about the so-called missile gap that ensued. The panic reestablished the power that scientists had garnered in the highest levels of policy-making during World War II. With renewed power, scientists "propounded the ideology of basic research" and secured for themselves more money with fewer strings attached (Geiger 1992, 43–44). Research monies in defense industries and defense projects flowed freely, driving up the bidding for scientific and technical talent and leaving industries and firms that did not participate in cold war research unable to compete for the best research talent. Contemporary accounts of this problem are legion. DuPont in the chemical industry and Xerox in copiers, for example, found themselves unable to hire the researchers they wanted.[70] The American utility industry was unable to continue to recruit high-caliber electrical engineers because work in electronics, computers, and space was far more appealing to students and their professors than was power engineering (Hirsh 1989).[71] Firms that had done commercially oriented R&D work historically but had then won sizable military contracts found themselves reallocating their scientific and technical talent. A two-class system (military and nonmilitary) ultimately developed, with the best and brightest concentrated in the military class.

As debates emerged over whether these effects were real or merely apparent, some analysts stressed the idea of spillovers, arguing that any R&D work done for the military with federal dollars would eventually be useful in commercial products. Others argued that the excessive military direction of R&D was undermining the long-run commercial health of the nation's economy. Robert Solo (1962) challenged the spillover argument in the pages of the *Harvard Business Review*, and Richard Nelson (1963) raised similar questions in an article in the *American Economic Review*. The 1965 *Report of the President's Committee on the Impact of Defense and Disarmament* contained corresponding analyses. This document, although accepting the idea of spillovers, argued that the same benefits could have been attained "at substantially lower costs and with more certainty" if research had been governed by market forces in the civilian sector.[72]

The role of military-related projects in funding industrial R&D became somewhat less dominant during the late 1960s and early 1970s, not because of concerns voiced by economists or the presidential commission, but because of the growing political costs of these policies. The tenor of the nation gradually shifted from being supportive of the national security state with all its scientific and technological trappings to being suspicious of it. There was a sense that the system had come to resemble George Orwell's *1984*. The Vietnam War, environmental concerns, urban crises, and the ho-hum public attitudes that beset NASA after the first few moon landings all contributed to the relative decline of federal support for industrial R&D (Geiger 1992, 45–47; Brooks 1986, 129; Morin 1993, 41; Mowery and Rosenberg 1989, 123–68; and Kevles 1978, 410–26).

This decline was attended by a shift in spending among many of the largest industrial R&D investors, including virtually all of the R&D pioneers. For almost two decades, firms had generously supported programs in fundamental research and managed fundamental researchers with kid gloves, as prescribed by the existing literature on the management of industrial research and development. By the late 1960s, however, executives in many of these firms had lost faith in Bush's linear model. They noted that few, if any, blockbuster products had emerged from their fundamental research programs. Although GE had produced some exotic things in its laboratory, the company had not earned much measurable return on its investment in academic-style research. DuPont had no new nylons. Kodak had no radically new system of photography. RCA had lost many opportunities, and one of its Princeton lab's products had failed to gain management support. Managers and executives at IBM began to question whether it had been wise to separate research from development; as one IBM researcher wrote, the move "had gone too far, . . . work in research was so esoteric that it no longer had value to IBM" (Pugh 1986, 567). Other corporate executives concluded that their firms had built ivory towers and, in so doing, neglected their older, existing businesses. These businesses were still producing the majority of sales but profit margins were now being squeezed. Firms that had dedicated major resources to new product R&D—even those that had put new products on the market—often found the products failing commercially or in serious need of additional research to make them profitable.[73] Thus, many corporations began to reallocate R&D monies from long-term to short-term objectives. Moreover, as inflation increased during the late 1960s and the first half of the 1970s, corporations held their overall R&D budgets flat, producing declines in R&D spending when counted in constant dollars. For the two generations of researchers who had grown up holding the values espoused in *Science—the Endless Frontier*, these changes produced enormous anxiety and malaise. The age of the great industrial R&D programs appeared to be waning rapidly, and images of twilight abounded in much of the literature on industrial R&D during this period. The shocks to the U.S. economy brought on by the OPEC oil embargo in the early 1970s simply compounded these problems and perceptions.

RESURGENCE OF RESEARCH: THE OLD OR THE NEW?

In the late 1970s and throughout much of the 1980s, there was a remarkable return to the pattern of the late 1950s and early 1960s. Science policy analyst Alexander Morin (1993) has mockingly called the revival of this pattern "The Return to Normalcy" (p. 44).[74] Federal spending for research performed by industrial firms once again went up, although it did not reach the heights it had attained in 1957. Moreover, industry expenditures on research rose by 14 percent per year during much of the first half of the 1980s and at a lower but still healthy rate during much of the second half (Morin 1993, p. 117). The defense buildup under President Ronald Reagan, symbolized most clearly by the Strategic Defense Initiative or "Star Wars" program, brought large-scale funding of military R&D to levels reminiscent of the earlier period. The buildup also directed 5 percent of these funds toward basic research—a percentage that some scholars have said is the price of university researchers' souls.[75]

Writing in the middle of the Reagan buildup, the historian Margaret Graham (1985a) stressed that "while the funding scenario may look like a return to the Age of Big Science, in other respects the current era for industrial research promises to be quite different." She went on to note the degree to which the federal government had abandoned the

Bush linear model by "encouraging cooperation in R&D and transfer between sectors, industries, and companies, trying to avoid the perceived problems of research competition that emerged in the [1950s and 1960s]" (p. 190). Even so, the swing toward military domination of the federal research budget, which Daniel Kevles termed "the re-militarization of American science," greatly worried some policymakers (Kevles 1987, 150). As director of the NSF from 1984 to 1990, Erich Bloch, a veteran of IBM development projects, launched an extensive series of engineering and interdisciplinary research centers dedicated to building knowledge and capabilities that cut across firms and industries and that were of direct commercial relevance. Bloch maintained that NSF's university-based engineering research centers were truly devoted to basic research unlike the basic research supported by mission-oriented agencies of the Department of Defense, NASA, and the Department of Energy. Research conducted by these agencies, Bloch and others argued, was conducted principally within industry, inherently biased toward the missions of those agencies, and therefore not really basic (Bloch 1986, 595–99).[76]

The resurgence of industry-funded R&D spending was attended by another recognizable trend: big spending by firms on university research. Monsanto's unprecedented $23 million research grant to Harvard University in 1974 was the opening move of several large corporate grants to major research universities. Exxon's 1980 grant of $8 million to MIT, DuPont's $6 million contract with Harvard Medical School in 1981, Hoechst's $50 million grant to Massachusetts General Hospital in 1981, and Mallinckrodt's $3.9 million contract with Washington University in 1981 suggested to some that a new trend was in the making. The agreements undertaken by these firms and others were part of a major strategy of firms, mostly in the chemical industry, to move their companies into biotechnology, the hottest area in industrial research from the late 1970s on. In some sense, the grants represented the price of admission to the biotechnology drama that was unfolding in university laboratories. Companies not only wanted access to the latest science, but they also hoped to learn who the best scientists were and then recruit them for their own programs. Many university scientists, however, shunned big firms and instead leveraged their university research to gain major equity in the numerous biotechnology start-up companies that dominated the scene.[77]

A major study, commissioned by the National Science Board, the governing body of the NSF, for inclusion in its 1982 annual report to the president and Congress, focused on these large industrial grants to universities and on less spectacular ones. The report examined the whole range of university-industry research relationships, including the NSF's formal grants program for university-industry research partnerships, an important prelude to Bloch's engineering research centers (National Science Foundation 1983). The study was the first of many that were done in the 1980s and 1990s to track the major trends of university-industry research relationships.[78] A 1994 report by Wesley Cohen, Richard Florida, and W. Richard Goe on some 1,100 university-industry research centers suggested that the federal government had been an active promoter of these centers (60 percent of which were formed in the 1980s) by tying federal contributions to universities to industry participation. These university-industry research centers represented roughly 70 percent of industry's expenditures on academic R&D. Of the $4.1 billion spent in the centers in 1991, local, state, and federal governments contributed 46 percent and the universities themselves 18 percent. Industry funded most of the remainder, with private foundations contributing a small fraction (Cohen, Florida, and Goe 1994).

Cooperation in the 1980s was not limited to the universities and industries. While pushing up federal spending for Star Wars and other defense-related research, the Reagan administration also began to dismantle much of the antitrust architecture that had shaped corporate strategies over the previous four decades. The administration's effort culminated in the National Cooperative Research Act of 1984, which paved the way for research consortia such as SEMATECH and the Microelectronic and Computer Technology Corporation (MCC) (Scott 1989, 65–84).[79] While these intraindustry cooperative research efforts, like the university-industry research centers, were ostensibly geared to shoring up or restoring American industry and research capabilities as global competition mounted, the government—more particularly, the Department of Defense—proved to be SEMATECH's major supporter and a large supporter of other such efforts. As the final crescendo in the cold war symphony was reached, economic security and military security remained inextricably bound together. Echoes of this cold war symphony were still resonating in the early 1990s and, thus, calls for a "civilian DARPA" (Defense Advanced Research Projects Administration) were hardly surprising.[80]

Concern about the competitiveness of the United States in the global economy and the natural desire to maintain many of the scientific and technical structures of the cold war led Congress to pass key pieces of legislation. Although the Stevenson-Wydler Technology Innovation Act of 1980 sought to transfer federally owned or developed technologies to state and local governments and, ostensibly, to the private sector, additional legislation was necessary to generate much enthusiasm for the process. The Federal Technology Transfer Act of 1986 and the National Competitiveness Technology Transfer Act of 1989 enabled national laboratories and federally funded research and development centers (contractor-operated laboratories) to transfer technologies that did not threaten national security out of the confines of those institutions and into the marketplace.[81] Agents for these organizations immediately began marketing their laboratories' knowledge, skill, hardware, and software. Many industrial firms, especially those with solid in-house capabilities, began shopping around these laboratories, looking for technological opportunities. Whether these technology transfer projects improved the nation's competitiveness or saved the national laboratories through what people called "defense conversion" after the Berlin Wall came down in 1990 and the Soviet Union collapsed in 1991 remained unclear (Adam 1990, 39–44; Council on Competitiveness 1992; Andrews 1993, C1, C4; and Office of Technology Assessment 1993). Conversion suggested to some, however, that the labor market for R&D personnel would be restructured over time, reversing some of the effects of the cold war buildup (Kilborn 1993, E3). Others maintained that it was only a matter of time before the government would begin to close some of the national laboratories bred by the cold war (David 1994, 3–8).[82]

The loosening of the antitrust statutes stimulated another development that affected the management of industrial R&D in the 1980s and early 1990s: joint ventures, or JVs. The 1980s and early 1990s witnessed a robust growth in the number of firms that undertook joint R&D projects, that is, projects in which in theory the R&D capabilities of two or more firms are combined and the costs and risks of the projects are shared. Whether this joint-venture phenomenon represented the release of a pent-up demand that had grown over many years of strict enforcement of antitrust statutes, a strong secular trend, or simply a fad remained unclear.[83]

This litany of developments, most of them initiated by the federal government but some supported vigorously by industry, should make clear that the 1980s and early 1990s brought about a crazy quilt of avenues, approaches, and opportunities for corporate R&D managers. The array of ways to spend money on R&D became staggering, especially in light of the increasing globalization of business, which had begun before the end of the cold war.

During the 1950s and early 1960s, many U.S. firms had established research laboratories in Europe in anticipation of or in response to the formation of the Common Market, which emerged from the Treaty of Rome in 1957. They did so for several different reasons, including the desire to create a means of recruiting highly qualified and talented European scientists and engineers; the need to provide U.S. corporations with the know-how to modify their products for European markets; the desire to open windows to European scientific developments; the hope of building bridges to the European universities; and the imperative to lower the costs of research. Of course, some U.S. corporations did not establish research organizations but rather inherited them when they acquired European companies (Ronstadt 1977). The negotiations that led to the Maastricht Treaty of December 1991 and the anticipated effects of a "United States of Europe" led to another wave of laboratory foundations or expansions in Europe by U.S. corporations. At the same time, the desire of U.S. corporations to penetrate Asian markets from Japan and South Korea to Malaysia and China, especially, resulted in a significant pattern of laboratory foundings in these markets in the late 1980s and early 1990s.[84]

These moves by U.S. firms were matched by the actions of Asian and European firms. In 1985, roughly two dozen R&D facilities built or acquired by Japanese corporations were in operation in the United States, often in high-technology centers such as California's Silicon Valley or along Route 128 near Boston. By 1992, this number was approaching 120 (Dalton and Serapio 1993).[85] Many Japanese firms also built research alliances with leading U.S. universities such as Stanford and MIT (Rudolph 1991, 69). German chemical, electrical, and electronics firms founded labs in the United States and aligned themselves with American universities; some of their U.S.-based research divisions were larger than those in Germany (Dalton and Serapio 1993). As biotechnology grew in importance, German chemical and pharmaceutical companies established biotechnology research laboratories in the United States because the increasing power of the ecologically motivated Greens political movement made such research difficult in their own country.

The globalization of industrial R&D has raised new questions about the management of research in the modern corporation. It has posed problems that transcend many of the problems that corporations faced when they established formal research programs in the early twentieth century. Dealing with different laws, social customs, educational systems, and work ethics has challenged research managers and corporations seeking global knowledge and power.[86] In the early 1990s, for example, the Eastman Kodak Company moved aggressively to establish industrial research programs in Japan. A change in the leadership of the corporation, however, led the company to abandon its Japanese lab, leaving its Japanese researchers, many of them "defectors" from Japanese companies, confused and angry (Pollack 1993). In late 1992 and early 1993, IBM cut $1 billion—roughly 20 percent—out of its annual R&D budget, a cut that had repercussions throughout Europe, Asia, and the United States (Sweet 1993, 75–79).

Globalization has also posed problems for those who are responsible for accounting for the research and development budget of the United States. The phenomenon simply adds fuel to the arguments made by Robert Reich in his paean to the global economy, *The Work of Nations* (1991). Detailing the global character of many products, Reich aptly titled a 1990 article for the *Harvard Business Review,* "Who is Us?" Nonetheless, believing that the United States needed to improve its competitiveness in the world economy, stave off the gains of foreign manufacturers in U.S. markets, and help American firms penetrate foreign markets, the Clinton administration committed itself in 1993 to pursuing a coordinated, effective industrial policy.[87] Administration policymakers were alarmed by the news that Japanese spending in R&D as a percentage of gross national product had surpassed that of the United States in 1986 and that the gap between the two countries had continued to widen in the 1990s. Moreover, when the calculations were run for *nondefense* R&D spending as a percentage of gross national product, both Japan and Germany had been spending considerably more than the United States since 1970: Japan almost 50 percent more in 1990, Germany roughly 30 percent more that same year (National Science Foundation 1993, ch. 4).

TOWARD A NEW EQUILIBRIUM?

The growth of industrial research has been one of the distinguishing features of the twentieth century. Firms that pioneered industrial R&D in the United States did so because they were threatened by competition, because they were engaged in a process of rationalizing their organizations (often in response to antitrust threats and actions), and because they saw benefits to internalizing R&D rather than relying on the market. Yet in internalizing these functions, the R&D pioneers and firms that emulated them were never completely free to do what they wanted vis-à-vis scientists and research-oriented engineers. Corporations were highly dependent on educational institutions and an elite cadre of university professors for their supply of researchers. These professors often had competing agendas and possessed ideals that were sometimes hostile to industrial R&D. Thus, industrial research was shaped by the American scientific community perhaps as much as or more than industrial research shaped the scientific community. Further influencing the course and nature of industrial research were the two world wars and the cold war. Seeking unrestricted, unquestioned, and unaccountable support for a scientific elite housed in a relatively few American universities (i.e., "best science"), the American scientific community used national security concerns to promulgate a linear model of technological development that rested squarely on basic research. The achievements of such wartime products as the atom bomb, radar, and the proximity fuse lent enormous credence to this model, which was adopted wholesale after World War II and quickened by the national security concerns that arose during the cold war. Thus, industrial research came to be skewed more and more toward basic research. It was pursued with the faith that blockbuster products would inevitably flow out of this research, whether funded by firms themselves or by the largess of the national security state.

This linear model began to fall apart in the mid-1960s, and reductions in industrial research set in. During the late 1970s, however, and after Ronald Reagan's election in 1980, concerns about the decline in U.S. competitiveness and the "Evil Empire" of the Soviet Union fueled a resurgence in industrial research funded both by industry and the military. Some scholars saw this resurgence as the remilitarization of research or a return to the normalcy of cold war research. Others saw it as providing new ways for firms to invest in research, whether through university research centers, joint ventures, private research laboratories, or research consortia. Perceived opportunities in biotechnology gave this era a different feel from that of the 1950s and added considerably to the dynamic growth of industrial research, especially after the late 1970s.

Growth in the diversity of research opportunities was one of the major developments of the 1980s and 1990s. After the disintegration of the Soviet Union, U.S. firms could even contract for research in the former Soviet Union for a fraction of what it would cost in the United States. Thus, industrial research has moved away from the hierarchies exploited by the R&D pioneers and toward markets in which science and technology can be bought, seemingly with few or no penalties or transaction costs.[88] This new flexibility in industrial research may be closely related to the secular movement toward what is called "flexible specialization" (Piore and Sabel 1984), which when logically played out means the downsizing of firms and the outsourcing of production (and research). On the other hand, these developments may be simply the logical outcomes of the end of the cold war and an apparent oversupply of highly trained scientists, many of whom have been unable to find positions in universities and governmental and industrial research labs (Kilborn 1983, E3; Browne 1994). Where and when a new equilibrium may be reached are unclear.[89] One thing is certain, however. The outcome will not be wholly determined either by American corporate decision makers or by global economic forces. If the past is any guide to the future, the scientific communities in the United States and the rest of the world will themselves have something to say about the conduct and future of industrial research.[90]

Notes

1. For the standard treatment of the scientific/utilitarian ideals of the American Philosophical Society, see Brooke Hindle (1956).

2. Some of the more interesting works include those by Howard R. Bartlett (1941, 19–77); Kendall Birr (1966; 1979); John Rae (1979); Leonard S. Reich (1985); and David C. Mowery and Nathan Rosenberg (1989, 21–58).

3. On the pure science ideal in American historiography, see Nathan Reingold (1972, 38–62). On professionalization of American science, see Nathan Reingold (1976, 33–69) and Sally Gregory Kohlstedt (1976).

4. In 1851, Alexander Dallas Bache gave a presidential address to the American Association for the Advancement of Science (AAAS) similar in tone and message to Rowland's, and Simon Newcomb made similar remarks to the AAAS in 1876. Rowland delivered an encore to his 1883 polemic in his presidential address at the second annual meeting of the American Physical Society in 1899. Entitled "The Highest Aim of the Physicist," the address is reprinted in Rowland (1902). While making pleas for pure science and touting pure science as the highest aim of the physicist, Rowland also played an important role in the establishment of the discipline of electrical engineering. On this, see Robert Rosenberg (1983).

5. "Best science" is the term used by Daniel Kevles to denote the kind of science idealized and advocated by American scientists like Henry Rowland who dreamed of a scientific elite that could perform whatever research they wished with unlimited funding at leading institutions and with complete autonomy and no accountability. See Kevles (1978, 53).

6. On the development of research universities, see Roger L. Geiger (1986) and Laurence Veysey (1965).

7. James Brittain (1971) has noted how at the turn of the century AT&T's executives purchased Michael Pupin's patents on the loading coil rather than backing the strong legal position of one of their own technical staff

members because they maintained a belief in the heroic inventor. In the first decade of this century, many of DuPont's executives also held similar beliefs (Hounshell and Smith 1988, 35–39).

8. On this point, see Steven W. Usselman (1991).

9. Compare Whitehead's statement with that of Norbert Wiener as quoted in David Noble (1977): "Edison's greatest invention was that of the industrial research laboratory. . . . The GE Company, the Westinghouse interests and the Bell Telephone Labs followed in his footsteps. . . ." (p. 113). For a slightly different version, see Norbert Wiener (1993, 65).

   Western Union's president, William Orton, may have believed that Edison had founded his laboratory to invent exclusively for Western Union. Orton had agreed that Western Union would pay Edison $100 per week for the "payment of laboratory expenses incurred in perfecting inventions applicable to land lines of telegraph or cables within the United States," but Edison sold his inventions to other companies as well and did not view the payment as an exclusive retainer (Israel 1989, 71).

10. For the best account of this development, see Robert Friedel and Paul Israel (1986).

11. On the creation of the West Orange laboratory and its subsequent history, see W. Bernard Carlson (1991a); and Andre Millard (1990). Edison built two important businesses out of his work at West Orange—the phonograph industry and the rudimentary beginnings of the motion picture industry. These achievements are noteworthy and were largely ignored by historians of industrial R&D until Millard published his book. Steven Usselman (1992) draws an important contrast between Thomas Edison and George Westinghouse.

12. By "research system," I mean something comparable to what is explored systematically in Richard Nelson (1993).

13. On the German organic chemicals industry and R&D, see John J. Beer (1959; 1958); and Georg Meyer-Thurow (1982). On the German electrical industry, see Paul Erker (1990).

14. On the development of two of the most important of these research institutes, see David Cahan (1988; 1982) and (on the Kaiser Wilhelm Society for the Advancement of the Sciences and its chemical research institutes) Jeffrey A. Johnson (1990b). It should be noted that German industry and industrialists were important sources of support for the German research institutes.

15. The growth and concentration of the German chemical industry, however, led to the domination of chemistry as a profession, by industry (Johnson 1990a).

16. David Mowery has reported considerably more industrial R&D laboratories in existence in the United States ca. 1900 than the handful of firms I have suggested here (Mowery 1981, 51–58; Mowery and Rosenberg 1989, 62–63). I am not sure how he arrived at his pre-1921 figures, but I suspect

that he included such laboratories as analytical and control laboratories, which I would not classify as industrial R&D laboratories. Arthur D. Little (1913), who did much to promote U.S. industrial R&D, stressed that the R&D pioneers I have enumerated were exemplary. He also noted only about fifty "other notable laboratories engaged in industrial research" (p. 799).

17. General Electric represented the J. P. Morgan-brokered merger (1892) of the Edison General Electric Company (formed in 1889 as the result of reorganizing the disparate Edison companies) and the Thomson-Houston Company. Edison General Electric had remained wedded to direct current electricity, while Thomson-Houston had developed considerable expertise in the technically superior alternating current, a technology that was dominated by Westinghouse Electric. The merger allowed General Electric to go head-to-head with Westinghouse. With Thomson-Houston, GE acquired the considerable scientific and technical talents of Elihu Thomson, one of the founders of Thomson-Houston and an active consultant to GE, who maintained his own laboratory in Lynn, Massachusetts, much in the way Edison had his in New Jersey. Thomson, who was far more of a corporate player than Edison, was widely respected by academic scientists for his scientific talents (Carlson 1991b). As a consultant, however, Thomson was not a director of industrial R&D for GE in the 1890s, nor did he possess the mathematical engineering brilliance of GE's Charles Steinmetz.

18. On Steinmetz's role in the creation of GERL, see Leonard S. Reich (1985, 64–67); George Wise (1985b, 75–80); and Ronald R. Kline (1992, 128–33).

19. On Whitney's graduate education in Germany, see George Wise (1985b, 41–46).

20. Whitney and his teacher Albert A. Noyes, who would become one of the most notable physical chemists in the United States and build both MIT's and Cal Tech's fortunes in chemistry, developed an important recovery process for a printing company that would earn them a substantial income. On Whitney's commercial interests in chemistry, see George Wise (1985b, 58–62). On Noyes, see John W. Servos (1990, 100–55, 251–98).

21. Note the lengths to which he went in recruiting William Coolidge to GERL. Coolidge had earlier declined Whitney's offer (Wise 1985b, 119–26).

22. On Langmuir and his work at GE, see George Wise (1985b, 149–61) and Leonard S. Reich (1983).

23. The final breakdown came after the laboratory had gone through a relatively long dry spell in terms of its easily noticeable contributions to the corporation's welfare. Wise discusses Whitney's breakdowns with a true biographer's skill (1985b, 127–31, 282–310).

24. The distinguished Nobel Prize-winning physicist of the University of Chicago, Albert Michelson, was upset when Jewett, his student, entered industrial research rather than staying in academia. In his study of Alfred Vail and AT&T, Louis Galambos (1992) differs with Reich and argues that Vail's

strategy for "universal service" was the principal factor motivating the establishment of the company's research program.

25. George Eastman wrote in a directive dated 1896: "I have come to think that the maintenance of a lead in the apparatus trade will depend greatly upon a rapid succession of changes and improvements, and with that aim in view, I propose to organize the Experimental Department in the Camera Works and raise it to a high degree of efficiency. If we can get out improved goods every year nobody will be able to follow us and compete with us. The only way to compete with us will be to get out original goods the same as we do" (Jenkins 1975, 184).

26. By "rationalization," I mean the concentration of production in a firm's most efficient plants (thereby closing down duplicate and less efficient plants) and the pursuit of greater productive efficiencies through the achievement of maximum economies of scale and the optimal allocation of corporate resources.

27. This book was reissued in a second edition (with John A. Leermakers, Mees's assistant) by the same company in 1950. For more information on Kodak R&D, see Jeffrey L. Sturchio (1985) and Reese V. Jenkins (1975, 300–39). Mees also introduced the innovation of the published paper series, *Abridged Scientific Publications*, which was widely copied by other major R&D firms (David A. Hounshell and John Kenly Smith 1988, 676, note 54).

28. See also the perceptive analysis of Louis Galambos (1979) and Leonard S. Reich (1985, 239–57).

29. See David A. Hounshell and John Kenly Smith, Jr. (1988) for an extensive treatment of the variety of roles that R&D played in the diversification of DuPont. On the concept of organizational capabilities, see Alfred D. Chandler (1990, 14–46) and William Lazonick, William Mass, and Jonathan West (1995). Wesley M. Cohen and Daniel A. Levinthal (1990) have developed the general concept of organizational capabilities into the concept of "absorptive capacity" vis-à-vis industrial research and development. See also Nathan Rosenberg (1990).

30. In 1944, when DuPont was seeking to hire temporarily physicists from the Metallurgical Laboratory at Chicago to help with the start up of the Hanford plutonium reactors that it had designed and built (based upon Chicago's suggestions and knowledge base), the company met with complete failure. Explaining why, Chicago physicist Samuel Allison, Arthur Compton's assistant, wrote to DuPont's Roger Williams on March 13, 1944, that the physicists at Chicago feared that any formal association with the company would ruin their chances of "obtaining an academic position" after the war. Physicist I. I. Rabi, who played an instrumental role in the radiation laboratory at MIT during World War II, once compared the linking of scientists to technological development to hitching a race horse with a draft horse; the scientist (race horse) who devotes himself or herself to technological development (work horse) fails to be a scientist after a while. See I. I. Rabi (1965, 10).

31. On Conant, see James Hershberg (1993).

32. On the history of A. D. Little, see E. J. Kahn (1986); Earl Place Stevenson (1953); and Richard L. Lesher (1963, chap. 3). See also Harold Vagtborg (1975). Lesher and Vagtborg also discuss the history of other independent research organizations.

33. Duncan had promoted the idea through his publications, such as *The Chemistry of Commerce* (1907). On the history of the Mellon Institute, see John W. Servos (1994). For information on industrial fellowships, see Arnold Thackray (1983, 219) and David A. Hounshell and John Kenly Smith, Jr. (1988, 290).

34. Little believed that American trade associations were the logical agencies for carrying out industrial research and that they had shirked their duties (1913, 799–800). In 1917, A. P. M. Fleming conducted a study of industrial research in the United States for the British government's Department of Scientific and Industrial Research in which he listed thirteen trade associations in the United States undertaking research work. Fleming was obviously guided in his survey by Little's earlier assessment of industrial research in the United States.

35. The impact of World War I on the American chemical industry is treated by David A. Hounshell and John Kenly Smith, Jr. (1988, 76–97); Arnold Thackray (1983, 216 ff.); Daniel P. Jones (1969); and Williams Haynes (1945, vol. 2). I have also learned a great deal about this period of the chemical industry's history from Kathryn Steen (1995). On Salvarsan and its synthesis and production in the United States during the war, see Patricia S. Ward (1981).

36. The history of R&D in the pharmaceutical industry is treated by Jonathan Liebenau (1987); John P. Swann (1988); and John Parascandola (1992; 1985; 1983).

37. See also the interpretation of Millikan's role by Robert H. Kargon (1982, 82–121).

38. Willis Whitney was also a member the Naval Consulting Board owing to Edison's respect for him and Whitney's stature as a corporate research manager. On the history of the NCB, see Daniel J. Kevles (1978, 106–38 passim); Thomas P. Hughes (1971, 243–74); and George Wise (1985b, 169, 182–94, 196).

39. NACA was created in 1915 but did not receive much funding until 1916 (Kevles 1978, 104–05). For a history of NACA, see Alex Roland (1985). Contrast Roland's interpretation of NACA research with that of David C. Mowery and Nathan Rosenberg (1989, 182). Walter Vincenti (1990) presents a deeply informed view of some of NACA's research.

40. GE's Whitney was especially important in this role, as discussed in George Wise (1985b, 203–06, 260–64). DuPont's Charles Reese would go on to be elected president of both the American Chemical Society and the American Institute of Chemical Engineers and the first leader of the Directors of

Industrial Research (DIR). Kodak's Kenneth Mees would also become an ambassador for industrial research, and AT&T's Frank Jewett would later be interpreted as being the feudal lord over one of the "scientific estates" in the United States and a critical voice in the scientific and technical mobilization for World War II.

41. On Westinghouse research, see Ronald Kline (1986). This program resulted in extensive scientific publication by the research staff recruited by Mees's protégé, including the future Nobel Prize winner Arthur H. Compton. But the program was essentially disbanded in 1920 by Westinghouse managers who, perhaps because they were members of an entrenched engineering culture, did not share the same enthusiasm for basic research. Years later, DuPont's Crawford H. Greenewalt, a champion of fundamental industrial research, would write in his Manhattan Project diary on December 28, 1942, that Compton held "peculiar ideas as to the difference between 'scientific' and 'industrial' research." Compton's experience at Westinghouse surely shaped those ideas, which Greenewalt (who believed there was no difference) sought to change. For the context of Greenewalt's remarks, see David A. Hounshell (1992).

42. Hale to Edward M. House, October 31, 1916, as quoted in Ronald Tobey (1971, 41); Robert A. Millikan address delivered at the University of Chicago, as quoted in Kargon (1982, 90).

43. The history of the National Research Endowment is treated by Daniel J. Kevles (1978, 185–87) and Lance E. Davis and Daniel J. Kevles (1974).

44. The history of the Industrial Research Institute is given in its annual report for 1993.

45. The U.S. steel industry as a whole did very little industrial research before World War I, and much of what it did do was forced upon it by its consumers and by the American Society of Testing Materials (Knoedler 1993b; 1993a; 1991).

46. U.S. Steel was one of the very few corporations that pledged money to the ill-fated National Research Fund. Herbert Hoover might have made the same statement about the steel company's motives as he attributed to the National Electric Light Association: "Don't fool yourself that they care a damn for pure science. What they want is to get into their reports, which will soon be examined by the Federal Trade Commission, that they are giving money for pure science research" (Kevles 1978, 187). See Kevles's general treatment of this era in his chapter, "Making the Peaks Higher." John W. Servos (1990, 202–50) provides a perspective on research in physical chemistry in the iron and steel industry.

47. Minutes of the DIR at the Hagley Museum and Library make clear how niggardly U.S. Steel was in setting up the new research division and the difficulties Johnston faced in getting anyone's attention in the company.

48. Davisson and Germer's prize-winning paper was "Diffraction of Electrons by a Crystal of Nickel," *Physical Review* 30 (1927): 705–40. For a refreshing

view on the importance of basic research in industry, see Nathan Rosenberg (1990).

49. DuPont's executives overlooked many of the shortcomings of the other fundamental research groups and failed to appreciate the very long-term and hard-to-measure payoffs of the fundamental research group in chemical engineering. On fundamental research in chemical engineering, see David A. Hounshell and John Kenly Smith, Jr. (1988, 275–85).

50. On the public image of scientists after the war, see Kenneth MacDonald Jones (1975).

51. Note that these shibboleths were essentially the same ones that Hale, Millikan, and others had propounded after World War I, as reflected in Robert Yerkes (1920). The Westinghouse Electric and Manufacturing Company anticipated the shift toward pure research in industry and government when it announced in late 1937 the establishment of the Westinghouse Research Fellowship program and the hiring of Edward U. Condon, a physicist at Princeton University, as associate director of its research laboratories. Under Condon's direction, Westinghouse assembled one of the most impressive programs of fundamental research in high energy physics in the United States, much of it built around a huge Van de Graaf generator. On the program's establishment, see *Science* (1937) and C. C. Furnas (1948, 304). On Condon's approach to pure science in industry, in which he linked the R&D programs of World War II to the attitudinal developments of the late 1930s, see Edward U. Condon (1942).

52. For an analysis of the role of engineering know-how and corporate capabilities in the Manhattan Project, see David A. Hounshell (1992); Richard G. Hewlett (1976); and Richard G. Hewlett and Oscar E. Anderson, Jr. (1962).

53. For a brief summary of what the American automobile industry alone contributed to World War II mobilization and production of war matériel, see James J. Flink (1988, 272–76).

54. A more recent overview is Ronald Kline (1987). See also Larry Owens (1994). Despite the triumphs of the scientists during the war, there was great controversy over control of R&D insofar as it related to war production. On this point, see Peter Neushul (1993, chap. 3). See also the history of synthetic rubber R&D and production during the war as discussed by Peter J. T. Morris (1989) and Peter Neushul (1993, chap. 4).

55. Americans' ideas about the frontier had been shaped by Frederick Jackson Turner and his frontier school of American history, which developed after Turner delivered and published his classic 1893 essay, "The Significance of the Frontier in American History." On Turner and his influence, see Howard R. Lamar (1969). Harvey Brooks (1986), one of the nation's leading figures in science policy, has pointed out that Bush's report effectively offered the nation a "new social contract," which "Today [March 1985], nearly 40 years after the publication of the Bush report, . . . remains remarkably intact, despite numerous alarums and excursions whose rhetoric

has generally outrun their practical effect" (p. 126). Alexander Morin (1993) has called the Bush report "the Magna Charta for U.S. science" (p. 19).

56. Daniel J. Kevles (1978) notes, however, that after Charles E. (Engine Charlie) Wilson became secretary of defense in the 1950s, military support for basic science began to wane. Kevles quotes Wilson as saying "Basic research is when you don't know what you are doing" (p. 383). Wilson, the former chairman of the board at General Motors, shared the views about pure research that had governed Charles Kettering, creator and long-time director of General Motors Research. Kettering did not believe in pure science. Not until the Soviets launched Sputnik did military budgets for pure science reach pre–1950 levels. Wilson then became an object of blame for his policies on basic research funding. On Kettering, see Stuart W. Leslie (1983).

57. On the purity of pure science for the military, see the provocative article by Stanton A. Glantz and Norm V. Albers (1974). In the realm of industrial research management, the navy supported the research and writing of an extensive textbook on financial control of industrial research by Robert Anthony, a noted accounting and business control systems educator at the Harvard Business School. See Robert N. Anthony (1952).

58. On the reorganization of Ford Motor Company into a modern organization, see Allan Nevins and Frank Ernest Hill (1962). In 1953, Ford mounted an exhibit on its R&D program in the company's visitor's center as part of its fiftieth anniversary celebration. In four months of that year alone, 600,000 visitors toured this exhibit, where they saw many wonders of modern science and industry. Earlier that year, President Dwight David Eisenhower had dedicated Ford's new Research and Engineering Center amid much fanfare (Nevins and Hill 1962, 373).

59. Data on the growth of research laboratories derives from comparisons of the 1946, 1950, and 1960 editions of *Industrial Research Laboratories of the United States*. The rush to build new laboratories drove up the costs of those laboratories at rates above the average building cost inflation experienced after the war. See David A. Hounshell and John Kenly Smith, Jr. (1988, 355–56). For a contemporary view of the trend toward fundamental research in industry, see Frank B. Jewett (1947, 16–20).

60. The company had also carried out an extensive amount of other wartime R&D for the Office of Scientific Research and Development and was, in fact, one of OSRD's largest industrial contractors (behind Western Electric, General Electric, and RCA and ahead of Westinghouse, Remington Rand, Monsanto, and Kodak). See Larry Owens (1994).

61. Stine had been elected to the executive committee in 1930. Although a big promoter of fundamental research, he expressed his skepticism about the expansion of this type of research in the industrial departments in correspondence with Crawford H. Greenewalt, February 8, 1945, and in the meeting of the Executive Committee, June 13, 1945. Stine reached the mandatory retirement age of sixty-five shortly after this battle.

62. Margaret Graham has observed the effects of antitrust prosecutions on R&D in her studies of RCA and Alcoa. See Graham (1986) and Graham and Pruitt (1990). General Electric, Kodak, and RCA, among other major firms, bought deeply into the linear model. See George Wise (1985a; 1984); Jeffrey L. Sturchio (1985); and Margaret B. W. Graham (1985b; 1986, 48–75).

63. See, for example, Richard R. Nelson (1962, 549–83). Only later was an alternative history of the transistor constructed, and this was done by those who had lost faith in the linear model. See, for example, M. Gibbons and C. Johnson (1970); J. Languish et al. (1972); and Charles Weiner (1973). These studies emerged after the Department of Defense published the results of Project Hindsight, which seriously challenged the linear model and set off a firestorm of debate on the science-technology relationship and the significance of basic research. On Project Hindsight, see Chalmers W. Sherwin and Raymond S. Isenson (1967). Indeed, the history of the transistor became a battleground over what was the "true" relationship between science and technology. The NSF countered the Project Hindsight report with a study carried out by the Illinois Institute of Technology Research Institute (1968).

64. A delegation from IBM visited Westinghouse in 1955 and noted the distinguished work being done there in basic science. The delegation also gathered that Westinghouse's fundamental research group had not contributed anything directly to the company. When queried about this, the leader of the Westinghouse program replied that his organization had a broader mandate—to benefit society in general—and that the rest of the Westinghouse company could partake of those benefits, just like anyone else. Such an attitude was not uncommon among those pursuing basic research in industry in the 1950s and early 1960s. As Pugh (1986) notes, however, "Westinghouse had apparently tried to correct this situation by placing research in the engineering organization, only to cause many of its scientists to leave" (p. 544).

65. In addition, the Monsanto Corporation took over the management of the Oak Ridge National Laboratory for a short time after the war, but the company met with problems in running the facility and withdrew from its agreement when the AEC would not agree to Monsanto's bid to have a high-flux reactor built in Saint Louis or Dayton (two of Monsanto's research sites) instead of Oak Ridge. AT&T became the contracting manager of Sandia National Laboratories (a division of Los Alamos National Laboratories) in 1949 when the regents of the University of California (operator of Los Alamos) decided that the university should disassociate itself from the actual production of atomic weapons. Bell Labs bore much of the brunt of this new contract in that many of its best research managers became managers at Sandia. For information on the work of AT&T, GE, DuPont, Monsanto, and other firms in the nation's atomic energy program, see Richard G. Hewlett and Oscar E. Anderson, Jr. (1962); Richard G. Hewlett and Francis Duncan (1972); Richard G. Hewlett and Jack M. Holl (1989); and Necah Stewart Furman (1990).

66. Simon Ramo (1988) presents a highly stylized account of an insider's role in several important cold war projects, including the ICBM. Nevertheless, his views on how staid General Electric had become in its research and how dynamic R&D was in the early cold war defense industry are well worth pondering.

67. The full history of cold war science and technology and the histories of many federally funded defense projects have not been written. Some of the relevant sources include Daniel J. Kevles (1978); Joan Lisa Bromberg (1983); Roger L. Geiger (1992, 1993); Harvey Sapolsky (1990); Stuart W. Leslie (1993); Clarence G. Lasby (1971); Jeffrey Stine (1986); Peter Galison and Bruce Hevly (1992); Peter Galison (1988); Daniel Kevles (1988); Paul K. Hoch (1988); Henry Etzkowitz (1988); S. S. Schweber (1988); Joan Lisa Bromberg (1991); Robert Seidel (1987); and Bruce L. R. Smith (1990). Harvey Brooks (1986) provides a good overview from the vantage point of a science policy analyst.

    For more on the space program, see also Vernon Van Dyke (1964); John Logsdon (1970); David H. Devorkin (1992); Howard E. McCurdy (1993); and Alan J. Levine (1994).

68. For a picture of the government's financing of industrial R&D, see Graham (1985b, Table 3, p. 57). Both Graham (1986) and George Wise (1985a) note how RCA and GE, respectively, came to be highly dependent on federal monies (mostly military) for their research. RCA and GE were by no means the exceptions. AT&T's research and development budgets in the early 1950s were swollen with military monies, as were Westinghouse's. During the late 1950s and early 1960s, almost 40 percent of Bell Laboratories' personnel were assigned to military programs (Fagen 1978, 356).

69. For the history of what was once the dominant firm in the industry— RCA—see Margaret B. W. Graham (1985b).

70. Seymour Melman (1965) discusses these problems. See also the excerpt, "The Negative Consequences of Defense Spending," in James L. Clayton (1970, 82–92). Margaret B. W. Graham (1985b) also analyzes the problem.

71. Hirsh's conclusions about engineering talent in the utility industry accord with my personal experience as an electrical engineering student in the late 1960s and early 1970s. At that time, power engineering was barely taught and that which was offered was poor in comparison to electronic circuits classes.

72. Chapter 1 of this report appears in James L. Clayton (1970). Quotation given here is from p. 57. President Lyndon Johnson became an increasingly vocal critic of government-sponsored basic research and issued a call in 1966 for increased relevance in government-sponsored research. See Daniel S. Greenberg (1966). See also Harry G. Johnson (1965, 127–36).

73. These problems and perceptions are treated in George Wise (1985a); David A. Hounshell and John Kenly Smith, Jr. (1988, 509–40, 573–89); Jeffrey L. Sturchio (1985); and Margaret B. W. Graham (1986, 220–35; 1985a). In a

1965 article entitled "Harnessing the R&D Monster" (p. 160) *Fortune* quoted Monsanto's Charles A. Thomas, who had built his career at Monsanto on research, as saying that "the nation's R&D is now stumbling in a plethora of projects, sinking in a sea of money, and is being built on a quicksand of changing objectives." As Gordon Moore makes clear in his contribution to this volume, he and Robert Noyce, seizing the opportunity to begin with a blank slate when organizing their new start-up company Intel in 1968, rejected the linear model under which they had operated at Fairchild. They determined that they would not have a separate R&D organization because they had encountered so many problems with transferring knowledge and technology from R&D to production at Fairchild (even though they were responsible for the organization at Fairchild). Instead, they built the R&D function into Intel's normal design and manufacturing operations. See chapter 7.

74. Viewing it from a different angle, Harvey Brooks (1986) calls it "The Period of Emphasis on Innovation Policy" (p. 132).

75. Foreman (1987) takes the strongest position on the basic research portion of military development projects.

76. Even before assuming the directorship of the NSF, Bloch had worked to build research cooperatives with the goal of generating research that could cut across disciplines and provide basic knowledge for industry. See Erich Bloch and James D. Meindl (1983, 237–42).

77. An early but solid analysis of the biotechnology initiatives is Susan Wright (1986).

78. My research shows at least eight other studies carried out and published between 1983 and 1992.

79. Multifirm cooperation in research had proven to be a major boon to development during World War II when antitrust statutes were suspended to deal with such research problems as catalytic cracking to raise output of aviation fuel. See the comments of Ernest W. Thiele on the results of Recommendation 41 of the U.S. Petroleum Administrator in Burton H. David and William P. Hellinger (1983, 176). On Sematech, see Peter Grindley, David C. Mowery, and Brian Silverman (1994). On MCC, see David V. Gibson and Everett M. Rogers (1994).

80. The calls for a civilian DARPA are reviewed by Otis L. Graham, Jr. (1992, 207–39). DARPA has, in fact, become a central player in the Clinton Administration's technology policy. See U.S. House of Representatives, Office of Technology Assessment (1993, 27, 28–29, 38, 69–70, 121–43). John A. Alic et al. (1992) call into question the use of DARPA (or ARPA) as an effective implement for commercial technology policy. Harvey M. Sapolsky (1994) suggests that post-cold war science policy is becoming more subject to political winds and cruder forms of pork barreling.

81. Alan Schriesheim (1990) provides a reasonable review of these various acts. See also Bernard Cole (1992).

82. As this article went to press, the U.S. House of Representatives and Senate were negotiating the nation's R&D budget for FY1997 and FY1998. It was clear that several of the federal laboratories would be closed "permanently," but the question of exactly which ones was subject to hourly change and intense speculation. The latest findings on the national laboratories are reported in National Academy of Sciences (1995).

83. The literature on joint ventures in R&D is enormous, episodic, and largely inadequate, but see Herbert I. Fusfeld and Carmela S. Haklisch (1985); William G. Ouchi and Michele Kremen Bolton (1988); and David C. Mowery (1989).

84. Data on U.S. industry's overseas R&D is reported in National Science Foundation (1993, 123–24). See also Louis Uchitelle (1989); Susan Moffat (1991); Mark Casson (1991); Ove Granstrand, Lars Håkanson, and Sören Sjölander (1992); John Dioso and David Hunter (1992).

85. Richard Florida and Martin Kenney (1994) provide a spatial analysis of Dalton and Serapio's data. They find that Japanese-owned R&D facilities in 1992 were heavily concentrated in California (51.3 percent), Michigan (9.6 percent), New Jersey (6.4 percent), and Massachusetts (4.5 percent). See also Evan Herbert (1989) and Bob Johnstone (1992).

86. I am indebted to Francis Honn, former research director at BASF and Henkel, for his enormous insights on the problem of managing a global R&D program owing to differences in both legal and social structures in different countries. A contrarian view on globalization of R&D can be found in Pari Patel and Keith Pavitt (1991). See also John Dunning (1994).

87. See William J. Clinton and Albert Gore, Jr. (1993). On August 4, 1994, the Clinton administration produced its policy statement "Science in the National Interest" (SNI), which deals with scientific research in general as a way of building the competitive position of the United States. See Colleen Cordes (1994).

88. The emphasis here is on the word *seemingly* because, as far as I know, no one has done a thorough, systematic transactions cost study of the increasing externalization of research that has occurred since 1980.

89. The lack of clarity in how deeply the end of the cold war would affect the scientific and technological institutions (universities, national laboratories, defense contractors, and, more generally, industrial research) that fed off it is evident in National Science Foundation (1993); U.S. Congress, Office of Technology Assessment (1993); Edward R. David, Jr. (1994); and Linda Geppert (1994). However, the National Academy of Sciences' (1995) latest study of federal funding of science and technology signals a major retreat from the dominant post-World War II science policies epitomized by Bush (1945).

90. Calls by university-based scientists at the 1994 spring meeting of the American Association of the Advancement of Science for the closure of numerous national laboratories (principally weapons labs) and the reallo-

cation of monies to increasingly squeezed university research portends the future.

References

Adam, John A. 1990. "Federal Laboratories Meet the Marketplace." *IEEE Spectrum* 28 (October 1990): 39–44.

Alic, John A., et al. 1992. *Beyond Spinoff: Military and Commercial Technologies in a Changing World.* Boston: Harvard Business School Press.

Allison, David K. 1981. *New Eye for the Navy: The Origins of Radar at the Naval Research Laboratory.* Washington, D.C.: U.S. Government Printing Office.

Andrews, Edmund L. 1993. "Swords to Plowshares: The Bureaucratic Snags." *New York Times,* 16, February.

Anthony, Robert N. 1952. *Management Controls in Industrial Research Organizations.* Boston: Graduate School of Business Administration.

Bartlett, Howard R. 1941. "The Development of Industrial Research in the United States." In *Research—A National Resource.* Report of the National Research Council to the National Resources Planning Board, December 1940. Washington, D.C.: U.S. Government Printing Office.

Baxter, James P. 1968. Reprint. *Scientists Against Time.* (Cambridge: MIT Press. Original edition, Boston: Little Brown, 1946.

Beard, Edmund. 1976. *Developing the ICBM: A Study in Bureaucratic Politics.* New York: Columbia University Press.

Beer, John J. 1958. "Coal Tar Dye Manufacture and the Origins of the Modern Industrial Research Laboratory." *Isis* 49: 123–31.

———. 1959. *The Emergence of the German Dye Industry.* Urbana: University of Illinois Press.

Birr, Kendall. 1966. "Science in American Industry." In *Science and Society in the United States,* edited by David D. Van Tassel and Michael G. Hall. Homewood, Ill.: Dorsey Press.

———. 1979. "Industrial Research Laboratories." In *The Sciences in the American Context: New Perspectives,* edited by Nathan Reingold. Washington, D.C.: Smithsonian Institution Press.

Bloch, Erich. 1986. "Basic Research and Economic Health: The Coming Challenge." *Science* 232: 595–99.

Bloch, Erich, and James D. Meindl. 1983. "Some Perspectives from the Field." In *University-Industry Research Relationships: Selected Studies,* edited by National Science Board, National Science Foundation. Washington, D.C.: U.S. Government Printing Office.

Bolton, E. 1945. Records of E. I. du Pont de Nemours & Co. Series II, part 2, Box 832, Hagley Museum and Library, Wilmington, Delaware.

Breech, E. 1949. Research 1949, Accession 65–71, Box 46, Ford Industrial Archives, Ford Motor Company, Dearborn, Michigan.

Brittain, James E. 1971. "The Introduction of the Loading Coil: George A. Campbell and Michael I. Pupin." *Technology and Culture* 11: 36–57.

———. 1976. "C. P. Steinmetz and E. F. W. Alexanderson: Creative Engineering in a Corporate Setting." *IEEE Proceedings* 64 (1976): 1413–17.

Bromberg, Joan Lisa. 1983. *Fusion: Science, Politics, and the Invention of a New Energy Source.* Cambridge, Mass.: MIT Press.

———. 1991. *The Laser in America, 1950–1970.* Cambridge: MIT Press.

Brooks, Harvey. 1986. "National Science Policy and Technological Innovation." In *The Positive Sum Strategy: Harnessing Technology for Economic Growth,* edited by Ralph Landau and Nathan Rosenberg. Washington, D.C.: National Academy Press.

Browne, Malcolm W. 1994. "Cold War's End Clouds Research as Openings in Science Dwindle." *New York Times,* 20 February. 1: 1.

Bush, Vannevar. 1945. *Science—the Endless Frontier: A Report to the President on a Program for Postwar Scientific Research.* Washington, D.C.: U.S. Government Printing Office.

Cahan, David. 1982. "Werner Siemens and the Origin of the Physikalisch-Technische Reichsanstalt, 1872–1887." *Historical Studies in the Physical Sciences* 12: 253–83.

———. 1988. *An Institute for an Empire: The Physikalisch-Technische Reichsanstalt, 1871–1918.* Cambridge: Cambridge University Press, 1988.

Carlson, W. Bernard. 1991a. "Building Thomas Edison's Laboratory at West Orange, New Jersey." *History of Technology* 13: 150–67.

———. 1991b. *Innovation as a Social Process: Elihu Thomson and the Rise of General Electric, 1870–1900.* Cambridge: Cambridge University Press.

Carothers, Wallace H. 1929. "An Introduction to the General Theory of Condensation Polymers." *Journal of the American Chemical Society* 51: 2548–559.

———. 1931. "Polymerization." *Chemical Reviews* 8: 353–426.

———. 1932. "Fundamental Research in Organic Chemistry at the Experimental Station—A Review." Records of E. I. du Pont de Nemours & Co., Central Research and Development Department. Accession 1784, Box 16, Hagley Museum and Library, Wilmington, Delaware.

Casson, Mark, ed. 1991. *Global Research Strategy and International Competitiveness.* Cambridge, Mass.: B. Blackwell.

Chandler, Alfred D., Jr. 1977. *The Visible Hand: The Managerial Revolution in American Business.* Cambridge, Mass.: Harvard University Press.

———. 1990. *Scale and Scope: The Dynamics of Industrial Capitalism.* Cambridge, Mass.: Belknap Press.

Clayton, James L., ed. 1970. *The Economic Impact of the Cold War.* New York: Harcourt, Brace & World.

Clinton, William J., and Albert Gore, Jr. 1993. *Technology for America's Economic Growth: A New Direction to Build Economic Strength.* 22 February. Washington, D.C.: U.S. Government Printing Office.

Cohen, Wesley M., and Daniel A. Levinthal. 1990. "Absorptive Capacity: A New Perspective on Learning and Innovation." *Administrative Sciences Quarterly* 35: 128–52.

Cohen, Wesley, Richard Florida, and Richard Goe. 1994. *University-Industry Research Centers in the United States.* Pittsburgh: Carnegie Mellon University.

Cole, Bernard. 1992. "DOE Labs: Models for Tech Transfer." *IEEE Spectrum* 29 (December): 53–68.

Conant, James B. Papers. Harvard University Archives. Cambridge, Massachusetts.

Condon, Edward U. 1942. "Physics in Industry." *Science* 96: 172–74.

Cordes, Colleen. 1994. "The Highest Priority." *The Chronicle of Higher Education* (10 August): A21–A22.

Council on Competitiveness. 1992. *Industry as a Customer of the Federal Laboratories.* Washington, D.C.: Council on Competitiveness.

Dalton, Donald, and Manuel Serapio. 1993. *U. S. Research Facilities of Foreign Companies.* Washington, D.C.: Japan Technology Program, Technology Administration, U. S. Department of Commerce.

David, Burton H., and William P. Hellinger, eds. 1983. *Heterogeneous Catalysis: Selected American Histories.* ACS Symposium Series, 222. Washington, D.C.: American Chemical Society.

David, Edward R., Jr. 1994. "Science in the Post-Cold War Era." *The Bridge* 24 (Spring): 3–8.

Davis, Lance E., and Daniel J. Kevles. 1974. "The National Research Fund: A Case Study in the Industrial Support of Academic Science." *Minerva* 12: 213–20.

Davisson, C. J., and L. Germer. 1927. "Diffraction of Electrons by a Crystal of Nickel." *Physical Review* 30: 705–40.

Devorkin, David H. 1992. *Science with a Vengeance: How the Military Created U.S. Space Sciences after World War II.* New York: Springer-Verlag.

Dioso, John, and David Hunter. 1992. "Western Firms See Japanese R&D as Key to Success." *Chemical Week* 150, no. 20: 20.

Duncan, Robert Kennedy. 1907. *The Chemistry of Commerce.* New York: Harper and Bros.

Dunning, John. 1994. "Multinational Enterprises and Globalization of Innovatory Capacity." *Research Policy* 23: 67–88.

Dyer, Davis. 1993. "Necessity as the Mother of Convention: Developing the ICBM, 1954–1958." Paper presented at the Business History Conference, Annual Conference, 19 March, at Boston, Massachusetts.

Erker, Paul. 1990. "Die Verwissenschaftlichung der Industrie: Zur Geschichte der Industrieforschung in den Europäischen und Americkanischen Electrokonzernen 1890–1930." *Zeitschrift für Unternehmensgeschichte* 21: 73–94.

Fagen, M. D., ed. 1978. *A History of Engineering and Science in the Bell System: National Service in War and Peace (1925–1975)*. N.p.: Bell Telephone Laboratories, Inc.

Fleming, A. P. M. 1917. *Industrial Research in the United States*. London: HM Stationery Office.

Flink, James J. 1988. *The Automobile Age*. Cambridge: MIT Press.

Florida, Richard, and Martin Kenney. 1994. "The Globalization of Innovation: The Economic Geography of Japanese R&D Investment in the United States." *Economic Geography* 70: 344–69.

Ford Industrial Archives. 1951. Press releases from Ford news bureau, October 4. AIL-74-18056. Ford Motor Company. Dearborn, Michigan.

Forman, Paul. 1987. "Behind Quantum Electronics: National Security as [the] Basis for Physical Research in the United States." Part 1. *Historical Studies in the Physical and Biological Sciences* 19: 149–229.

Friedel, Robert, and Paul Israel. 1986. *Edison's Electric Light: Biography of an Invention*. New Brunswick: Rutgers University Press.

Furman, Necah Stewart. 1990. *Sandia National Laboratories: The Postwar Decade*. Albuquerque: University of New Mexico Press.

Furnas, C. C. ed. 1948. *Research in Industry: Its Organization and Management*. New York: D. Van Nostrand Company.

Fusfeld, Herbert I., and Carmela S. Haklisch. 1985. "Cooperative R&D for Competitors." *Harvard Business Review* (November–December): 4–11.

Galambos, Louis. 1979. "The American Economy and the Reorganization of the Sources of Knowledge." In *The Organization of Knowledge in Modern America, 1860–1920*, edited by Alexandra Oleson and John Voss. Baltimore: Johns Hopkins University Press.

———. 1992. "Theodore N. Vail and the Role of Innovation in the Modern Bell System." *Business History Review* 66 (Spring): 95–126.

Galison, Peter. 1988. "Physics between War and Peace." In *Science, Technology, and the Military*. Vol. 1, edited by Everett Mendelsohn, et al. Dordecht, The Netherlands: Kluwer.

Galison, Peter, and Bruce Hevly, eds. 1992. *Big Science: The Growth of Large-Scale Research*. Stanford: Stanford University Press.

Geiger, Roger L. 1986. *To Advance Knowledge: The Growth of American Research Universities, 1900–1940*. New York: Oxford University Press.

————. 1992. "Science, Universities, and National Defense, 1945–1970." *Osiris* 2d ser., 7: 26–48.

————. 1993. *Research and Relevant Knowledge: American Research Universities since World War II.* Oxford University Press.

Geppert, Linda. 1994. "Industrial R&D: The New Priorities." *IEEE Spectrum* 31 (September): 30–41.

Gibbons, M., and C. Johnson. 1970. "Relationship Between Science and Technology." *Nature* 227 (11 July): 125–27.

Gibson, David V., and Everett M. Rogers. 1994. *R&D Collaboration on Trial: The Microelectronics and Computer Corporation.* Boston, Mass.: Harvard Business School Press.

Glantz, Stanton A., and Norman V. Albers. 1974. "Department of Defense R&D in the University." *Science* 186: 706–11.

Gomory, Ralph. 1992. "The Technology-Product Relationship: Early and Late Stages." In *Technology and the Wealth of Nations*, edited by Nathan Rosenberg, Ralph Landau, and David Mowery. Stanford: Stanford University Press.

Goodstein, Judith R. 1991. *Millikan's School: A History of The California Institute of Technology.* New York: W. W. Norton.

Gorn, Michael H. 1988. *Harnessing the Genie: Science and Technology Forecasting for the Air Force, 1944–1986.* Washington, D.C.: Office of Air Force History.

Graham, Margaret B. W. 1985a. "Corporate Research And Development: The Latest Transformation." *Technology in Society* 7: 179–95.

————. 1985b. "Industrial Research in the Age of Big Science." *Research on Technological Innovation, Management and Policy* 2: 47–79.

————. 1986. *RCA & The VideoDisc: The Business of Research.* New York: Cambridge University Press.

Graham, Margaret B. W., and Bettye H. Pruitt. 1990. *R&D for Industry: A Century of Technical Innovation at Alcoa.* New York: Cambridge University Press.

Graham, Otis L., Jr. 1992. *Losing Time: The Industrial Policy Debate.* Cambridge: Harvard University Press.

Granstrand, Ove, Lars Håkanson, and Sören Sjölander, eds. 1992. *Technology Management and International Business.* New York: John Wiley & Sons.

Greenberg, Daniel S. 1966. "Basic Research: The Political Tides are Shifting." *Science* 152: 1724–726.

Greenewalt, Crawford H. Manhattan Project Diaries. Accession 1889. Hagley Museum and Library, Wilmington, Delaware.

————. Papers. Accession 1814, Box 37. Hagley Museum and Library, Wilmington, Delaware.

Grindley, P., D. C. Mowery, B. Silverman. 1994. "Sematech and Collaborative Research: Lessons in the Design of High-Technology Consortia." *Journal of Policy Analysis and Management* 13: 723–58.

Haynes, Williams. 1945. *The American Chemical Industry.* Vol. 2. New York: Van Nostrand.

Herbert, Evan. 1989. "Japanese R&D in the United States." *Research Technology Management* 32 (November–December): 11–20.

Hershberg, James. 1993. *James B. Conant: Harvard to Hiroshima and the Making of the Nuclear Age.* New York: Knopf.

Hewlett, Richard G. 1976. "Beginnings of Development in Nuclear Technology." *Technology and Culture* 17: 465–78.

Hewlett, Richard G., and Oscar E. Anderson, Jr. 1962. *The New World: A History of the United States Atomic Energy Commission, 1939–1946.* University Park: Pennsylvania State University Press.

Hewlett, Richard G., and Francis Duncan. 1972. *Atomic Shield: A History of the United States Atomic Energy Commission.* Washington, D.C.: U.S. Atomic Energy Commission.

———. 1974. *The Nuclear Navy, 1946–1962.* Chicago: University of Chicago Press.

Hewlett, Richard G., and Jack M. Holl. 1989. *Atoms for Peace and War, 1953–1961: Eisenhower and the Atomic Energy Commission.* Berkeley: University of California Press.

Hindle, Brooke. 1956. *The Pursuit of Science in Revolutionary America.* Chapel Hill: University of North Carolina Press.

Hirsh, Richard. 1989. *Technology and Transformation in the Electric Utility Industry.* New York: Cambridge University Press.

Hoch, Paul K. 1988. "The Crystallization of a Strategic Alliance: The American Physics Elite and the Military in the 1940s." In *Science, Technology, and the Military.* Vol. 1, edited by Everett Mendelsohn, et al. Dordecht, The Netherlands: Kluwer.

Hoover, Herbert. 1926. "The Vital Need for Greater Financial Support to Pure Science Research." *Mechanical Engineering* 48 (January): 6–8.

Hounshell, David A. 1980. "Edison and the Pure Science Ideal in 19th-Century America." *Science* 207: 612–17.

———. 1989. "The Modernity of Menlo Park." In *Working at Inventing: Thomas A. Edison and the Menlo Park Experience,* edited by William S. Pretzer. Dearborn, Mich.: Henry Ford Museum and Greenfield Village.

———. 1992. "Du Pont and the Management of Large-Scale Research and Development." In *Big Science: The Growth of Large-Scale Research,* edited by Peter Galison and Bruce Hevly. Stanford: Stanford University Press.

Hounshell, David A., and John Kenly Smith, Jr. 1988. *Science and Corporate Strategy: Du Pont R&D, 1902–1980*. Cambridge: Cambridge University Press.

Hughes, Thomas P. 1971. *Elmer Sperry: Inventor and Engineer*. Baltimore: The Johns Hopkins University Press.

Illinois Institute of Technology Research Institute. 1968. *Technology in Restrospect and Critical Events in Science*. Chicago: Illinois Institute of Technology.

Industrial Research Institute. 1994. *Annual Report—1993*. Washington, D.C.: Industrial Research Institute.

Israel, Paul. 1989. "Telegraphy and Edison's Invention Factory." In *Working at Inventing: Thomas A. Edison and the Menlo Park Experience*, edited by William S. Pretzer. Dearborn, Mich.: Henry Ford Museum and Greenfield Village.

———. 1992. *From Machine Shop to Inventor*. Baltimore: The Johns Hopkins University Press.

Jenkins, Reese V. 1975. *Images and Enterprise: Technology and the American Photographic Industry, 1839–1925*. Baltimore: The Johns Hopkins University Press.

Jewett, Frank B. 1947. "The Future of Scientific Research in the Postwar Era." In *Science in Progress*, no. 5, edited by George A. Baitsell. New Haven: Yale University Press.

Johnson, Harry G. 1965. "Federal Support of Basic Research: Some Economic Issues." In *Basic Research and National Goals: A Report to the Committee on Science and Astronautics of the U.S. House of Representatives*. Washington, D.C.: National Academy of Sciences.

Johnson, Jeffrey A. 1990a. "Academic, Proletarian, . . . Professional? Shaping Professionalization for German Industrial Chemists, 1887–1920." In *German Professions, 1800–1950*, edited by Geoffrey Cocks and Konrad H. Jarausch. New York: Oxford University Press.

———. 1990b. *The Kaiser's Chemists: Science and Modernization in Imperial Germany*. Chapel Hill: University of North Carolina Press.

Johnstone, Robert. 1992. "Research: Setting Up on Enemy Ground." *Far Eastern Economic Review* 155, no. 24 (June 18): 54–5.

Jones, Daniel P. 1969. "The Role of Chemists in War Gas Research in the United States during World War I." Ph.D. diss., University of Wisconsin.

Jones, Kenneth MacDonald. 1975. "Science, Scientists, and Americans: Images of Science and the Formation of Federal Science Policy, 1945–1950." Ph.D. diss., Cornell University.

Kahn, E. J. 1986. *The Problem Solvers: A History of Arthur D. Little, Inc.* Boston: Little, Brown.

Kargon, Robert H. 1982. *The Rise of Robert Millikan: Portrait of a Life in American Science*. Ithaca, New York: Cornell University Press.

Kay, Herbert, 1965. "Harnessing the R&D Monster." *Fortune* (January): 160–163, 196–198.

Kevles, Daniel J. 1977. "The National Science Foundation and the Debate over Postwar Research Policy, 1942–1945: A Political Interpretation of *Science— The Endless Frontier.*" *Isis* 68: 5–26.

———. 1978. *The Physicists: The History of a Scientific Community in Modern America*. New York: Knopf.

———. 1987. "The Remilitarization of American Science: Historical Reflections." Manuscript. As quoted in Paul Foreman, "Behind Quantum Electronics: National Security as [the] Basis for Physical Research in the United States." Part 1. *Historical Studies in the Physical and Biological Sciences* 19: 149–229.

———. 1988. "An Analytical Look at R&D and the Arms Race." In *Science, Technology, and the Military*. Vol. 2, edited by Everett Mendelsohn, et al. The Netherlands: Kluwer.

———. 1990. "Principles and Politics in Federal R&D Policy, 1945–1990: An Appreciation of the Bush Report." Introduction to *Science—The Endless Frontier* by Vannevar Bush. Reprint edition. Washington, D.C.: National Science Foundation.

Kilborn, Peter T. 1993. "The Ph.D.'s Are Here, But the Lab Isn't Hiring," *New York Times*, 18 July, E3.

Kline, Ronald. 1986. "The Origins of Industrial Research at the Westinghouse Electric Company, 1886–1922." Paper presented at the Annual Meeting, Society for the History of Technology, 25 October. Pittsburgh.

———. 1987. "R&D: Organizing for War." *IEEE Spectrum* 24 (November): 54–60.

———. 1992. *Steinmetz: Engineer and Socialist*. Baltimore: The Johns Hopkins University Press.

Knoedler, Janet T. 1991. "Backward Linkages to Industrial Research in Steel, 1870–1930." Ph.D. diss., University of Tennessee.

———. 1993a. "Early Examples of User-Based Industrial Research." *Business and Economic History* 22 (Fall): 285–94.

———. 1993b. "Market Structure, Industrial Research, and Consumers of Innovation: Forging Backward Linkages to Research in the Turn-of-the-Century U.S. Steel Industry." *Business History Review* 67 (Spring): 98–139.

Kohlstedt, Sally Gregory. 1976. *The Formation of the American Scientific Community*. Urbana: University of Illinois Press.

Komons, Nick A. 1966. *Science and the Air Force: A History of the Air Force Office of Scientific Research*. Arlington, Virginia: Office of Aerospace Research.

Koppes, Clayton R. 1982. *JPL and the American Space Program: A History of the Jet Propulsion Laboratory*. New Haven: Yale University Press.

Lamar, Howard R. 1969. "Frederick Jackson Turner." In *Pastmasters*, edited by Marcus Cunliffe and Robin Winks. New York: Harper & Row.

Langrish, J., M. Gibbons, C. Johnson, and F. R. Jevons. 1972. *Wealth from Knowledge*. New York: Halsted/John Wiley.

Lasby, Clarence G. 1971. *Project Paperclip: German Scientists and the Cold War*. New York: Atheneum.

Lazonick, William, William Mass, and Jonathan West. 1995. "Strategy, Structure, and Performance: Comparative-Historical Foundations of the Theory of Competitive Advantage." In *Proceedings of the Conference on Business History*, October 24 and 25, 1994, Rotterdam, The Netherlands, edited by Mila Davids, Ferry de Goey, and Dirk de Wit. Rotterdam: Centre of Business History, Erasmus University.

Lesher, Richard L. 1963. "Independent Research Institutes and Industrial Application of Aerospace Research." Ph.D. diss., Indiana University.

Leslie, Stuart W. 1983. *Boss Kettering: Wizard of General Motors*. New York: Columbia University Press.

———. 1993. *The Cold War and American Science*. New York: Columbia University Press.

Levine, Alan J. 1994. *The Missile and Space Race*. Greenwood, Conn.: Praeger.

Levine, Arnold S. 1982. *Managing NASA in the Apollo Era*. Washington, D.C.: National Aeronautics and Space Administration.

Liebenau, Jonathan. 1987. *Medical Science and Medical Industry: The Formation of the American Pharmaceutical Industry*. Baltimore: The Johns Hopkins University Press.

Little, Arthur D. 1913. "Industrial Research in America." *Journal of Industrial and Engineering Chemistry* 5: 793–801.

Logsdon, John. 1970. *The Decision to Go to the Moon: Project Apollo and the National Interest*. Cambridge: MIT Press.

McCurdy, Howard E. 1993. *High Technology and Organizational Change in the American Space Program*. Baltimore: Johns Hopkins University Press.

McDougall, Walter. 1985. *The Heavens and the Earth: A Political History of the Space Age*. New York: Basic Books.

Mees, C. E. Kenneth. 1916. "Organization of Industrial Research Laboratories." *Science* 43: 763–73.

———. 1920. *The Organization of Industrial Scientific Research*, New York: McGraw-Hill.

Melman, Seymour. 1965. *Our Depleted Society*. New York: Holt, Rinehart and Winston.

Merck & Co., Inc. 1991. *Values and Visions: A Merck Century*. Rahway, N.J.: Merck & Co., Inc.

Meyer-Thurow, Georg. 1982. "The Industrialization of Invention: A Case Study from the German Chemical Industry," *Isis* 73: 363–81.

Millard, Andre. 1990. *Edison and the Business of Innovation*. Baltimore: The Johns Hopkins University Press.

Millikan, Robert A. 1982. "The New Opportunity in Science." *Science* 50 (1919): 285–297. As quoted in Robert H. Kargon, *The Rise of Robert Millikan: Portrait of a Life in American Science*. Ithaca, New York: Cornell University Press.

Moffat, Susan. 1991. "Picking Japan's Research Brains." *Fortune* 123 (25 March): 85–96.

Morin, Alexander J. 1993. *Science Policy and Politics*. Englewood Cliffs, N.J.: Prentice-Hall.

Morris, Peter J.T. *The American Synthetic Rubber Research Program*. Philadelphia: University of Pennsylvania Press, 1989.

Mowery, David C. 1981. "The Emergence and Growth of Industrial Research in American Manufacturing, 1899–1945," Ph.D. diss., Stanford University.

———. 1983. "The Relationship between Contractual and Intrafirm Forms of Industrial Research in American Manufacturing, 1900–1944." *Explorations in Economic History* 20 (October): 351–74.

———. 1989. "Collaborative Ventures between U.S. and Foreign Manufacturing Firms." *Research Policy* 18: 19–32.

Mowery, David C., and Nathan Rosenberg. 1989. *Technology and the Pursuit of Economic Growth*. New York: Cambridge University Press.

National Academy of Sciences, Committee on Criteria for Federal Support on Research and Development, 1995. *Allocating Federal Funds for Science and Technology*. Washington, D.C.: National Academy of Sciences.

National Resources Planning Board. 1938. *Research—A National Resource* Vol. 1. Washington, D.C.: U.S. Government Printing Office.

National Science Foundation. National Science Board. 1983. *University-Industry Research Relationships: Selected Studies*. Washington, D.C.: U.S. Government Printing Office.

National Science Foundation. National Science Board. 1993. *Science and Engineering Indicators—1993*. Washington, D.C.: U.S. Government Printing Office.

Nelson, Richard R. 1959. "The Economics of Invention: A Survey of the Literature," *The Journal of Business* 32 (April): 101–27.

———. 1962. "The Link Between Science and Invention: The Case of the Transistor." In *The Rate and Direction of Inventive Activity: Economic and Social Factors*, edited by Richard R. Nelson. Princeton: Princeton University Press.

————. 1963. "The Impact of Arms Reduction on Research and Development." *American Economic Review* 53: 435–46.

————, ed. 1993. *National Innovation Systems: A Comparative Analysis.* New York: Oxford University.

Neushul, Peter. 1993. "Science, Technology, and the Arsenal of Democracy: Production Research and Development during World War II." Doctoral diss., University of California at Santa Barbara.

Nevins, Allan and Frank Ernest Hill. 1962. *Ford: Decline and Rebirth, 1933–1962.* New York: Charles Scribner's Sons.

Noble, David. 1977. *America by Design: Science, Technology and the Rise of Corporate Capitalism.* New York: Knopf.

Ouchi, William G., and Michele Kremen Bolton. 1988. "The Logic of Joint Research and Development." *California Management Review* (Spring): 9–33.

Owens, Larry. "The Counterproductive Management of Science in the Second World War: Vannevar Bush and the Office of Scientific Research and Development." *Business History Review* 68 (Winter): 515–76.

Parascandola, John. 1983. "Charles Holmes Herty and the Effort to Establish an Institute for Drug Research in Post World War I America." In *Chemistry and Modern Society,* edited by John Parascandola and James C. Wharton. Washington, D.C.: American Chemical Society.

————. 1985. "Industrial Research Comes of Age: The American Pharmaceutical Industry, 1920–1940." *Pharmacy in History* 27: 12–21.

————. 1990. "The 'Preposterous Provision': The American Society for Pharmacology and Experimental Therapeutics' Ban on Industrial Pharmacologists, 1908–1941." In *Pill Peddlers: Essays on the History of the Pharmaceutical Industry,* edited by Jonathan Liebenau et al. Madison: American Institute of the History of Pharmacy.

————. 1992. *The Development of American Pharmacology: John J. Abel and the Shaping of a Discipline.* Baltimore: The Johns Hopkins University Press.

Patel, Pari, and Keith Pavitt. 1991. "Large Firms in the Production of the World's Technology: An Important Case of 'Non-Globalization.'" *Journal of International Business Studies* 22: 1–21.

Piore, Michael, and Charles Sabel. 1984. *The Second Industrial Divide: Possibilities for Prosperity.* New York: Basic Books.

Pollack, Andrew. 1993. "Cutbacks at Kodak Lab in Japan Create Unease." *New York Times,* 15 February, D2.

Pugh, Emerson W. 1986. "Research." In *IBM's Early Computers,* edited by Charles J. Bashe, Lyle R. Johnson, John H. Palmer, and Emerson W. Pugh. Cambridge: MIT Press.

Rabi, I. I. 1965. "The Interaction of Science and Technology." In *The Impact of Science on Technology*, edited by Aaron W. Warner, Dean Morse, and Alfred S. Eichner. New York: Columbia University Press.

Rae, John. 1979. "The Application of Science to Industry." In *The Organization of Knowledge in Modern America, 1860–1920*, edited by Alexandra Oleson and John Voss. Baltimore: The Johns Hopkins University Press.

Ramo, Simon. 1988. *The Business of Science: Winning and Losing in the High-Tech Age*. New York: Hill and Wang.

Records of E. I. du Pont de Nemours & Co. Hagley Museum and Library, Wilmington, Delaware.

Redmond, Kent C., and Thomas M. Smith. 1980. *Project Whirlwind: The History of a Pioneer Computer*. Bedford, Mass.: Digital Press.

Reich, Leonard S. 1983. "Irving Langmuir and the Pursuit of Science and Technology in the Corporate Environment." *Technology and Culture* 24: 199–221.

———. 1985. *The Making of American Industrial Research: Science and Business at GE and Bell, 1876–1926*. New York: Cambridge University Press.

Reich, Robert B. 1990. "Who Is Us?" *Harvard Business Review* (January–February): 53–64.

———. 1991. *The Work of Nations: Preparing Ourselves for 21st-Century Capitalism*. New York: Knopf.

Reingold, Nathan. 1972. "American Indifference to Basic Research." In *Nineteenth-Century American Science: A Reappraisal*, edited by George H. Daniels. Evanston, Ill.: Northwestern University Press.

———. 1976. "Definitions and Speculations: The Professionalization of Science in America in the Nineteenth Century." In *The Pursuit of Knowledge in the Early American Republic: American Scientific and Learned Societies from the Colonial Times to the Civil War*, ed. Alexandra Oleson and Sanborn C. Brown. Baltimore: Johns Hopkins University Press.

Report of the President's Committee on the Impact of Defense and Disarmament. 1965. Washington, D.C.: U.S. Government Printing Office.

Roland, Alex. 1985. *Model Research: The National Advisory Committee for Aeronautics, 1915–1958*. 2 vols. Washington, D.C.: National Aeronautics and Space Administration.

Ronstadt, Robert. 1977. *Research and Development Abroad by U.S. Multinationals*. New York: Praeger.

Rosenberg, Nathan. 1990. "Why Do Firms Do Basic Research (With Their Own Money)?" *Research Policy* 19: 165–74.

Rosenberg, Robert. 1983. "American Physics and the Origins of Electrical Engineering," *Physics Today* 36: 48–54.

Rowland, Henry A. 1902. *Physical Papers of Henry A. Rowland*. Baltimore: The Johns Hopkins University Press.

Rudolph, Barbara. 1991. "Follow That Brain Wave." *Time*, 12 August, 69.

Russo, Arturo. 1981. "Fundamental Research at Bell Laboratories: The Discovery of Electron Diffraction." *Historical Studies in the Physical Sciences* 12: 117–60.

Sapolsky, Harvey M. 1990. *Science and the Navy: The History of the Office of Naval Research*. Princeton: Princeton University Press.

———. 1994. "Financing Science After the Cold War." In *The Fragile Contract: University Science and the Federal Government*, edited by David H. Guston and Kenneth Kenniston. Cambridge: MIT Press.

Schriesheim, Alan. 1990. "Toward a Golden Age for Technology Transfer." *Issues in Science and Technology* 7, no. 2: 52–8.

Schweber, Silvan S. 1988. "The Mutual Embrace of Science and the Military: ONR and the Growth of Physics in the United States after World War II." In *Science, Technology, and the Military*. Vol. 1, edited by Everett Mendelsohn, et al. The Netherlands: Kluwer.

Science. 1937. "The Westinghouse Research Fellowship." 86 (31 December): 605–06.

Scott, John T. 1989. "Historical and Economic Perspectives of the National Cooperative Research Act." In *Cooperative Research and Development*, edited by Albert N. Link and Gregory Tassey. Boston: Kluwer.

Seidel, Robert W. 1983. "Accelerating Science: The Postwar Transformation of the Lawrence Radiation Laboratory." *Historical Studies in the Physical Sciences* 13: 375–400.

———. 1986. "A Home for Big Science: The Atomic Energy Commission's Laboratory System." *Historical Studies in the Physical and Biological Sciences* 16: 137–75.

———. 1987. "From Glow to Flow: A History of Military Laser Research and Development." *Historical Studies in the Physical and Biological Sciences* 18: 111–47.

———. 1990. "Clio and the Complex: Recent Historiography of Science and National Security." *Proceedings of the American Philosophical Society* 13: 420–41.

———. 1994. "Accelerators and National Security: The Evolution of Science Policy for High-Energy Physics, 1947–1967." *History and Technology* 11: 361–91.

Servos, John W. 1990. *Physical Chemistry from Ostwald to Pauling: The Making of a Science in America*. Princeton: Princeton University Press.

———. 1994. "Changing Partners: The Mellon Institute, Private Industry, and the Federal Patron," *Technology and Culture* 35 (April): 221–57.

Sherwin, Chalmers W., and Raymond S. Isenson. 1967. "Project Hindsight: A Defense Department Study of the Utility of Research." *Science* 156 (June 23): 1571–577.

Sigethy, Robert. 1980. "The Air Force Organization for Basic Research, 1945–1970: A Study in Change." Ph.D. diss., American University.

Smith, Bruce L.R. 1990. *American Science Policy Since World War II*. Washington, D.C.: Brookings Institution.

Solo, Robert A. 1962. "Gearing Military R&D to Economic Growth." *Harvard Business Review* (November–December): 49–60.

Steen, Kathryn. 1995. "Wartime Catalyst and Postwar Reaction: The Making of the U.S. Synthetic Organic Chemicals Industry, 1910–1930." Ph.D. diss., University of Delaware.

Stevenson, Earl Place. 1953. *Scatter Acorns That Oaks May Grow: Arthur D. Little, Inc., 1886–1953*. New York: Newcomen Society of North America.

Stine, Jeffrey. 1986. *History of U.S. Science Policy Since World War II: Report of the Task Force on Science Policy*. U.S. House of Representatives Committee on Science and Technology. Washington, D.C.: U.S. Government Printing Office.

Sturchio, Jeffrey L. 1985. "Experimenting with Research: Kenneth Mees, Eastman Kodak, and the Challenges of Diversification." Paper presented at the R&D Pioneers Conference, 7 October. Wilmington, Delaware.

Sturm, Thomas. 1967. *The USAF Scientific Advisory Board: Its First Twenty Years, 1944–1964*. Washington, D.C.: Office of Air Force History, U.S. Air Force.

Swann, John P. 1988. *Academic Scientists and the Pharmaceutical Industry: Cooperative Research in Twentieth-Century America*. Baltimore: The Johns Hopkins University Press.

Sweet, William. 1993. "IBM Cuts Research in Physical Sciences at Yorktown Heights and Almaden." *Physics Today* 46, no. 6: 75–9.

Thackray, Arnold. 1983. "University-Industry Connections and Chemical Research: An Historical Perspective." In *University-Industry Research Relationships: Selected Studies*, edited by National Science Board, National Science Foundation. Washington, D.C.: U.S. Government Printing Office.

Tiffany, Paul A. 1986. "Corporate Culture and Corporate Change: The Origins of Industrial Research at the United States Steel Corporation, 1901–1929." Paper presented at the Annual Meeting of the Society for the History of Technology, October 25, Pittsburgh.

Tobey, Ronald C. 1971. *The American Ideology of National Science, 1919–1930*. Pittsburgh: University of Pittsburgh Press.

Tocqueville, Alexis de. 1876. *Democracy in America*. Vol. 2. Boston: John Allyn.

Uchitelle, Louis. 1989. "U.S. Companies Lift R&D Abroad." *New York Times*, 22 February, D2.

U.S. Congress, Office of Technology Assessment. 1993. *Defense Conversion: Redirecting R&D*, OTA-ITE-552. Washington, D.C.: U.S. Government Printing Office.

Usselman, Steven W. 1985. "Running the Machine: The Management of Technological Innovation on American Railroads, 1860–1910," Ph.D. diss., University of Delaware.

———. 1991. "Patents Purloined: Railroads, Inventors, and the Diffusion of Innovation in 19th-Century America." *Technology and Culture* 32: 1047–75.

———. 1992. "From Novelty to Utility: George Westinghouse and the Business of Innovation during the Age of Edison." *Business History Review* 66: 251–304.

Vagtborg, Harold. 1975. *Research and American Industrial Development: A Bicentennial Look at the Contributions of Applied R&D*. New York: Pergamon Press.

Van Dyke, Vernon. 1964. *Pride and Power: The Rationale of the Space Program*. Urbana: University of Illinois Press.

Veysey, Laurence. 1965. *The Emergence of the American University*. Chicago: University of Chicago Press.

Vincenti, Walter G. 1990. *What Engineers Know and How They Know It*. Baltimore: The Johns Hopkins University Press.

Ward, Patricia S. 1981. "The American Reception to Salvarsan." *Journal of the History of Medicine* 36: 44–62.

Weart, Spencer. 1979. "The Physics Business in America, 1919–1940: A Statistical Reconnaissance." In *The Sciences in the American Context: New Perspectives*, edited by Nathan Reingold, Washington, D.C.: Smithsonian Institution Press.

Weiner, Charles. 1970. "Physics in the Great Depression." *Physics Today* 23 (October): 31–38.

———. 1973. "How the Transistor Emerged." *IEEE Spectrum* 10 (January): 24–33.

Whitehead, Alfred North. 1985. *Science and the Modern World*. London: Free Association Books. (Original edition, London: The Macmillan Company, 1926.)

Wiener, Norbert. 1993. *Invention: The Care and Feeding of Ideas*. Cambridge: MIT Press.

Wise, George. 1984. "Science at General Electric." *Physics Today* 37, no. 12: 52–61.

———. 1985a. "R&D at General Electric, 1878–1985." Paper presented at the R&D Pioneers Conference, 7 October, Wilmington, Delaware.

———. 1985b. *Willis R. Whitney, General Electric, and the Origins of U.S. Industrial Research*. New York: Columbia University Press.

Wright, Susan. 1986. "Recombinant DNA Technology and Its Social Transformation, 1972–1982." *Osiris,* 2d Ser., no. 2: 303–60.

Yerkes, Robert, ed. 1920. *The New World of Science: Its Development During the War.* New York: Century.

2

THE ROLES OF UNIVERSITIES IN THE ADVANCE OF INDUSTRIAL TECHNOLOGY

Nathan Rosenberg and Richard R. Nelson

THE LAST FIFTEEN YEARS have seen growing interest in two questions: How does university research relate to technical advance in industry; and how can American university research become a more effective contributor to the competitiveness of American industry?* American industry has had some thoughts about this matter and has put considerable monies behind them. Industry funding of academic research rose from 2.6 percent in 1970, to 3.9 percent in 1980, to an estimated 6.9 percent in 1990 (National Science Foundation 1991). In a recent study, Cohen, Florida, and Goe (1994) estimated that about 19 percent of university research now occurs in programs that are linked with industry in some essential manner.

The U.S. government has also given these questions considerable thought. A large part of the discrepancy between the National Science Foundation's figure of 6.9 percent in 1990 and Cohen, Florida, and Goe's figure of 19 percent can be explained by the fact that the federal government provides funds for programs that are linked to industry. In the last decade both Congress and the executive branch have come to believe that university research ought to be connected more closely than it has been with industry and ought to be more responsive to industry's needs. The engineering research centers sponsored by the National Science Foundation (NSF) are only a small part of the overall policy response to this belief. By and large, universities have welcomed

* The authors would like to acknowledge the valuable comments of L. E. Birdzell, Jr., Harvey Brooks, Michael Crow, Stephen Kline, Scott Stern, Walter Vincenti, and Robert White.

increased business funding for research and government support of university programs that are linked to industry. Administrators know that the argument that university research ought to be supported for reasons of national military security carries far less weight with the end of the cold war and that the argument that university research helps American competitiveness now has the most promise of maintaining the flow of government funds. Furthermore, many administrators who feel that government support for research has plateaued, see industry as a potential source of significant new funding.

It is striking, however, how little of the current discussion in industry, government, or in academia is informed by a careful examination of the roles that university research actually plays in industrial technical advance. In fact, these roles differ from industry to industry, and they have changed significantly over the years. The current debate is proceeding with little grounding in what is going on now, why it is going on, and how we got to where we now are. The principal purpose of this chapter is to provide this grounding. The first part of the chapter will provide a historical perspective. The second part will consider the division of labor that exists today between industrial research and academic research and the important differences in this division across industries and technologies. The final part will address the current debate.

A HISTORICAL PERSPECTIVE

From the early days of the republic until the end of World War II, American universities had a widespread reputation for being "practice and vocation" oriented. Alexis de Tocqueville commented on the attitudes toward science in the young republic in the 1830s:

> In America the purely practical part of science is admirably understood and careful attention is paid to the theoretical portion which is immediately requisite to application. On this head, the Americans always display a clear, free, original, and inventive power of mind. But hardly anyone in the United States devotes himself to the essentially theoretical and abstract portion of human knowledge. (Tocqueville 1876, 48)

In the 1800s and early 1900s, British visitors sneered at what they perceived to be the "vocationalism" of the American higher educational system. Educational institutions in the United States took on responsibilities for teaching and research in fields such as agriculture and mining, in commercial subjects such as accounting, finance, marketing and management, and in an ever-widening swath of engineering subjects—civil, mechanical, electrical, chemical—long before their British (or, for the most part, other European) counterparts did so.

The passage of the Morrill Act of 1862, which gave federal lands to states establishing colleges that offered programs in agriculture, engineering, and home economics, reflected the American view of the appropriate roles of university teaching and research at that time and provided substantial and steady support for these roles. The new legislation worked within and supported the tradition that was then developing of university responsiveness to the demands of the community. While the focus here is on university research, research and training have always been intertwined in the American higher educational system. A good portion of the research done at American universities prior to World War II reflects this intertwining. Thus, not only did the University of Akron supply skilled personnel for the local rubber industry, it also became well known for its research in the processing of rubber. While the land-grant colleges are rightly praised for fostering the high productivity of the American farm through the teaching of food production skills, practical solutions to agricultural problems emerged from research at these universities that was spurred by the specific needs of the local agricultural community. Because they were responsive to the needs of the community, state universities often developed programs addressing an extremely diverse range of needs. Thus, after World War I, the University of Illinois offered programs in architectural engineering, ceramic engineering, mining engineering, municipal and sanitary engineering, railway civil engineering, railway electrical engineering, and railway mechanical engineering. As one observer noted, "Nearly every industry and government agency in Illinois had its own department at the state university in Urbana—Champaign" (Levine 1986, 52).

Although usually connected with training, university research programs that were aimed to meet the needs of local industry often took on a life of their own and became quite institutionalized. The University of Oklahoma was long known for its research in the field of petroleum, while the University of Kentucky and the University of North Carolina worked extensively on developing technologies relating to the processing of tobacco. For many years, the University of Illinois and Purdue University worked on railroad technologies, ranging from the design of locomotive boilers to maintenance and repair. To this day, the Purdue football team is called the Boilermakers. Occasionally, university research on the technical problems of an industry involved large-scale, long-run commitments to the solution of a particular problem. The Mine Experimentation Station's taconite project at the University of Minnesota was directed at the problems caused by the gradual exhaustion of high-yielding iron ores in the Mesabi range. Research on how to use ore with a lower iron content and more impurities began before World War I and continued into the early 1960s. The various processing technologies developed by this project considerably lengthened the useful life of the Mesabi range (Davis 1964).

Of course, much of the early research to help local industry was idiosyncratic, and until the late nineteenth or early twentieth century, there was little in the way of a systematic disciplinary basis for this work. Thus, there was little to tie together the individuals and universities engaged in similar fields. In fact, one of the major accomplishments of American universities during the first half of the twentieth century was the institutionalization of new engineering and applied science disciplines. It was at this time that chemical engineering, electrical engineering, and aeronautical engineering became established fields of study in American universities. In each of these fields, programs of graduate studies with certified professional credentials grew up, professional organizations were founded, and associated journals came into existence. The rise of these new disciplines and programs was induced by and, in turn, supported the growing use of university-trained engineers and scientists in industry. At the same time, these new disciplines supported the rise of the industrial research laboratory, first in the chemical and electrical equipment industries and later throughout American industry.

Although the engineering disciplines draw from science and are applications oriented, it would be a mistake to interpret them only as vehicles for applying basic science to practical ends. Chemical engineering, for example, is not applied chemistry. It cannot be adequately characterized as the application of scientific knowledge generated in a chemical laboratory. Indeed, chemical engineering emerged precisely because the knowledge generated in chemistry fell far short of the kinds of knowledge needed to produce a new product on a commercial scale. Chemical engineering involves a merger of chemistry and mechanical engineering, i.e., the application of mechanical engineering to the large-scale production of chemical products. Moreover, chemical engineering could not emerge as an academic discipline until an entirely new methodology, totally distinct from the science of chemistry, was created. This new methodology, which exploited the concept of "unit operations" developed by Arthur D. Little in 1915, provided the basis for a rigorous, quantitative approach to large-scale chemical manufacturing and permitted the systematic quantitative instruction of future practitioners. It made chemical engineering a form of generic knowledge that could be taught at universities.

Like chemical engineering, electrical engineering is far more than "applied physics." Electrical engineering, like chemical engineering, involves a heavy element of mechanical engineering and also an analysis of the nature of complex systems. Also like chemical engineering, electrical engineering soon developed a research agenda of its own. That is to say electrical engineers in American universities did research in such specific areas as high voltage, network analysis, and the insulating properties of different materials and configurations.

These and other applications-oriented fields are complex areas of inquiry that require scientific training and sophistication for their effective exploration. They emerge as distinct fields after a technology and an industry come into existence. In the post–World War II era computer science emerged as an academic field after the modern computer. In some cases, new basic scientific understanding is won in the course of efforts to comprehend the behavior of a new technology. William Shockley's efforts to understand how transistors worked, for example, led him to develop the whole new subject of "imperfections in almost perfect crystals," a subject that was crucial to determining the influence of minority carriers in semiconductors. His research also led to many new developments in crystal physics, which later became important in metallurgy and eventually led to the emergence of the new applied discipline of materials science. In applications-oriented fields, then, the scientific research agenda is harnessed to the need for a better understanding of a technology. In these fields, the traditional distinctions between "basic" and "applied" sciences are anachronisms because these fields are, to use Herbert Simon's apt phrase, "sciences of the artificial" (Simon 1969).

In order to understand what research in modern universities is all about and the relationship of that research to industrial research, it is important to recognize and understand these "sciences of the artificial." Today, applications-oriented fields account for a large share, significantly more than half, of all academic research. By their very nature these fields are linked to the practical problems of industry, agriculture, medicine, and other areas of human concern. Thus, the widespread notion that academic research is about things far removed from practical concerns is, quite simply, mistaken. There is a similar misconception about the nature of "basic research." By general agreement, the term is reserved for research that involves the quest for fundamental understanding of natural phenomena. In the traditional natural sciences, this quest has often been identified with research that is significantly distanced from any concern about practical application. However, the widely accepted notion of basic research today has come to focus on the *absence* of a concern with practical applications, not on the search for fundamental understanding. This is not only unfortunate, but bizarre. In the applied sciences and engineering disciplines, some of the research done is basic research in the sense that it is a search for understanding at a fundamental level. Thus, medical studies of carcinogenic processes often involve basic research on cell biology. Research in the field of computer science may involve the development of new mathematics or an inquiry into the way humans solve problems. Indeed, some recent research on turbulence involves very fundamental issues. In all of these areas, however, the agenda for research is strongly influenced by the nature of real life problems.

By the start of World War II, the applied sciences and engineering disciplines were firmly established in the American university system. While a few ivy-league institutions, Harvard and Yale for example, tended to resist or to isolate them, they were strong at most of the land grant universities, which accounted for a large fraction of American university research. Their presence significantly influenced, but did not replace, the long-standing tradition of research in the service of local industry and agriculture and the training of people to enter industry.

World War II was a watershed in the history of American science and technology and led to drastic changes in the roles played by American universities in the nation's scientific and technical enterprises. During the war, most of the country's scientific and technical capabilities were mobilized to work on projects aimed at hastening the successful termination of the war. While the Manhattan Project, which developed the atomic bomb, was the most dramatic of these research endeavors and the one that most caught the imagination of the American people, there were many other projects, such as research on radar, synthetic rubber, and the proximity fuse. As a result, the prestige of American academic science grew enormously. While large-scale, federally supported university research was unthinkable prior to World War II, wartime successes made the unthinkable, thinkable. Federal funding of academic research, which probably amounted to no more than one-quarter of total academic research support in the mid-1930s, increased enormously. By 1960, it accounted for over 60 percent of all funding. Total academic research funding increased more than tenfold between 1935 and 1960 and more than doubled again by 1965 (see Table 2-1). Over this same period, the Consumer Price Index (CPI) increased more than twofold (from 41.1 in 1935 to 88.7 in 1960, where prices in 1967 = 100) and more than 6 percent between 1960 and 1965. While the CPI is not fully adequate as a research expense deflator, it is plausible that by 1965 real resources going into academic research were more than twelve times what they had been in the mid-1930s. Rapid growth continued from 1965 until 1980 or so. During this time, it is estimated that total real academic research funding grew at a rate of about 3 percent per year.

With the vast expansion of research funding and the federal government's growing role in this funding, there came a dramatic transformation in the character of university research. While it is true that American universities had begun to do world class research in astronomy and in certain areas of fundamental physics and chemistry in the years between World War I and World War II, there was now a major shift toward basic research in all academic science fields. Basic research

Table 2-1.   Support for Academic R&D by Sector
1935–1991
(Millions of Current Dollars)

Year	Total Academic R&D	Federally Supported R&D	Federal Percentage of Total
1935	$ 50	$ 12	24%
1960	646	405	63
1965	1,474	1,073	73
1970	2,335	1,647	71
1975	3,409	2,288	67
1980	6,077	4,104	68
1985	9,686	6,056	63
1990 (est.)	16,000	9,250	58

Sources: Data for 1935: National Resources Committee, *Research—A National Resource* (Washington, D.C.: U.S. Government Printing Office, 1938), 2:178. Data for 1960 and after: National Science Foundation, *Science and Engineering Indicators 1991* (Washington, D.C.: U.S. Government Printing Office, 1991), 348.

became not only respectable but also widely perceived as what universities ought to be doing. By the mid-1960s, the American system was clearly leading the way in most scientific fields. Statistics of Nobel prizes and publications in scientific journals tell part of the story, but the best indicator of the American system's success is the flow of students from Europe to the United States for graduate training, a reversal of the situation prior to the war.

Even as American universities became preeminent centers of basic research and graduate education, the dominant rationale for most federal research funding continued to be the expectation that research would yield practical benefits. While the NSF is indeed committed to basic research for its own sake, the agency has accounted for less than one-fifth of the federal support for university research since its creation after World War II. The Department of Defense and two other government agencies allied with it in many ways—NASA and the Department of Energy (previously the Atomic Energy Commission)—account for much more, roughly one quarter of the total. By far, the largest government funder of academic research by the mid-1970s, accounting for more than 45 percent of total university and academic research funding was the National Institutes of Health (NIH) (see Table 2-2).

The mission orientation of the biggest funders is reflected in the distribution of research funding by field. As shown in Table 2-3, research funding in the life sciences accounts for more than one-half of all funds for academic science research. Research funding in the engineering disciplines exceeds funding in the physical sciences.

While much of the research in these fields is basic research, and is so in the sense that researchers often are aiming for deep under-

Table 2-2. Percent of Federal Research Funds Originating Within Particular Agencies

Year	NIH	NSF	DOD	NASA	DOE	USDA	Other
1971	36.7	16.2	12.8	8.2	5.7	4.4	16.0
1976	46.4	17.1	9.4	4.7	5.7	4.7	12.0
1981	44.4	15.7	12.8	3.8	6.7	5.4	11.0
1986	46.4	15.1	16.7	3.9	5.3	4.2	8.4
1991 (est.)	47.2	16.1	11.6	5.8	4.7	4.0	10.7

Source: National Science Foundation, *Science and Engineering Indicators 1991* (Washington, D.C.: U.S. Government Printing Office, 1991), 360.

Table 2-3. Federal and Nonfederal R&D Expenditures at Universities and Colleges, by Field, 1989

Field	Thousands of Dollars	Percent
TOTAL SCIENCE & ENGINEERING	$14,987,279	100.0
TOTAL SCIENCES	12,599,686	84.1
Life Sciences	8,079,851	53.9
Physical Sciences	1,643,377	11.0
Environmental Sciences	982,937	6.6
Social Sciences	636,372	4.2
Computer Sciences	467,729	3.1
Psychology	237,945	1.6
Mathematical Sciences	214,248	1.4
Other Sciences	337,227	2.3
TOTAL ENGINEERING	2,387,593	15.9
Electrical/Electronic	600,016	4.0
Mechanical	340,280	2.3
Civil	249,552	1.7
Chemical	185,087	1.2
Aero/Astronautical	146,548	1.0
Other	866,110	5.8

Sources: National Science Foundation, *Academic Science/Engineering: R&D Expenditures, Fiscal Year 1989*, NSF 90-321, Detailed Statistical Tables (Washington, D.C.: U.S. Government Printing Office, 1991); unpublished tabulations.

standing, concern for practical application is very much in the minds of the funders and, in many cases, in the minds of the academics doing the research. The major role of American universities in the development of the modem computer is a story that has been told and told again. The close connection between university researchers and biotechnology companies today is well-known and serves as an outstanding example of basic research and practical application.

As a result of the changes that occurred after World War II, however, research aimed directly at helping local civilian industry and agriculture, which had once been the hallmark of American university research, waned. While many universities, for example RPI and Georgia Tech, did continue to help local industry, schools like MIT and Cal Tech tended to draw away from that function. Although federal and state funding for agricultural research actually increased, it became a relatively small part of total university research funding. By 1960, defense and health-related problems had become the dominant foci and the rationale for university research funding. While a portion of the research continued to be very much "hands-on, dirt-under-the-nails" kinds of work, the notion that academic research should lead to scientific breakthroughs became dominant. Conceptual generality became a principal criterion for "good" research, even in the engineering and applied science disciplines. Deborah Shapley and Rustum Roy (1985) comment constantly on this change in orientation and the low prestige of engineering in academia in relation to the prestige of the pure sciences. However, we believe that they overstate their case, because, as we have noted, whatever its standing in terms of prestige, engineering has received more research funding than the physical sciences, and research at medical schools has received far more funding than research conducted in the biology departments of liberal arts schools.

It was the rise of concerns about the competitiveness of American industry in the 1980s that rekindled notions of academic research in service to civilian industry. The end of the cold war and the erosion of the power of national security as a rationale for public support of universities has also led to a rethinking of old notions. In the concluding section of this chapter, we will offer our commentary on the current debate, but before doing this, it is important to look more closely at the roles American universities currently play in technical advance in industry.

THE CURRENT ROLES OF AMERICAN UNIVERSITY RESEARCH

In the last fifty years, there has developed a relatively clear division of labor between academic and industrial research. This division has been supported by the policies of the DOD, the NIH, and the NSF. In fields in which firms have strong R&D capabilities, R&D to improve existing products and processes or aimed at bringing into practice and commercial use the next generation of products and processes has become almost exclusively the province of industry. In a few industries, some firms do engage in long-term research oriented toward advancing understanding. For the most part, however, basic research in industry, although it accounts for more than one-fifth of all basic research in the United States, constitutes only 5 percent of all industrial R&D. Today, except for those fields in which, in effect, university work is substituting for industrial R&D, university research is basic research.

As we have already noted, this does not mean the work is not motivated by or funded because of its promise to deal with a class of practical problems, nor does it mean that university scientists and engineers are not building and working with prototypes of applicable industrial technology. Indeed, this is a central part of academic research in many engineering fields. What it does mean, however, is that cases like the taconite project at the University of Minnesota are unusual, and so too are those cases in which academic medical scientists carry their work into operational practice. Today, university research usually stimulates and enhances the power of R&D done in industry. Most of the work involved in creating and bringing into practice new industrial technology is carried out in industry, not in universities.

One good way of seeing what it is that universities do *not* do is to recognize that the bulk of the effort that goes into R&D in most technologies goes into D, not R. If we consider total R&D spending for the American economy, D has constituted approximately two-thirds of that total for many years. Except when special institutions or projects are established (the University of Minnesota's Mines Experiment Station, university-affiliated agricultural experiment stations, industry-servicing engineering facilities like those at Georgia Tech and RPI, or special DOD projects, for example), academic institutions are not motivated by or good at D. Usually, most of the science employed in achieving the objective of a marketable new technology is rather old science. This is not the kind of work that excites academics, and its successful completion usually does not lead to publication and tenure. Moreover, the understandings that are most important in guiding product and process development efforts are often those associated with detailed familiarity with prevailing technology and user needs rather than familiarity with the most recent research findings (Rosenberg 1982, ch. 6; von Hippel 1988). Although most universities are not set up to perform development work, university research has certainly contributed to the advance of industrial technology. The DOD and NIH's energetic buildup of the academic research enterprise in fields of particular interest to them led academics in these fields to develop many prototypes of new technology that were subsequently developed in industry. In fact, on some occasions academics have been involved in development work as well.

A survey of industrial R&D managers, undertaken in the mid-1980s by one of the authors of this article and several of his colleagues at Yale, provides a wealth of data that reveal the industrial fields in which university research is most important and makes it possible to see how university research contributes to the advance of industrial technology. Respondents were asked to rate the importance of research done at universities to technical advance in their industry. A striking number of the industries that rated university research as highly important were related to agriculture or forestry. This clearly reflects the long-standing service research role of universities, especially state universities, for industries that provide key inputs for agriculture or that process agricultural or forest products. While this type of service research has been dwarfed by university research funded by such agencies as the DOD and NIH, it is apparent that it continues to be critical. When the significant university role in such fields as plant patents is considered, it becomes evident that this research is a substitute for, not a complement to, private research. The electronics industries, the scientific and measurement instrument industries, and the pharmaceutical industries also rated university research as very important. This clearly reflects the prominence of DOD and NIH funding of university research.

Respondents were also asked to rate on a scale from 1 to 6, the relevance of university research performed in specific fields. Table 2-4 shows the number of industries that gave various fields of university research a high relevance score. It is striking that a large fraction of the research rated as important by a number of industries comes from the applied science or engineering fields. Very little of the research performed by the basic sciences is mentioned with the exception of chemistry. Chemistry is the exception because a significant fraction of the research is done with a deep appreciation of practical industry problems. Thus, in some cases, as in the research on catalysis, such work may win a Nobel Prize in chemistry and also contribute to the ability of chemical companies to produce more effectively.

The fact that few industries in Table 2-4 find research in such fields as physics and mathematics relevant does not mean that academic research in these fields contributes little to technical advance. It takes a long time for fundamental advances in physics, mathematics, and similar sciences to have an impact on industrial technology, and in our view, that impact, when it occurs, tends to be indirect. Thus, advances

Table 2-4. The Relevance of University Research in a Specific Field to Industrial Technology

Field	Number of Industries Giving University Research in the Field a High Relevance Score		Selected Industries in Which the Relevance of Research Is High
	5 or 6	*6*	
Biology	12	3	Animal feed, drugs, processed fruits and vegetables
Chemistry	19	3	Animal feed, meat products, drugs
Geology	0	0	None
Mathematics	5	1	Optical instruments
Physics	4	2	Optical instruments, electron tubes
Agricultural Science	17	7	Pesticides, animal feed, fertilizers, food products
Applied Math/ Operations Research	16	2	Meat products, logging/sawmills
Computer Science	34	10	Optical instruments, logging/sawmills, paper machinery
Materials Science	29	8	Synthetic rubber, nonferrous metals
Medical Science	7	3	Surgical/medical instruments, drugs
Metallurgy	21	6	Nonferrous metals, fabricated metal products
Chemical Engineering	19	6	Canned foods, fertilizers, malt beverages
Electrical Engineering	22	2	Semiconductors, scientific instruments
Mechanical Engineering	28	9	Hand tools, specialized industrial machinery

Source: Previously unpublished data from Yale Survey on Appropriability and Technological Opportunity, Yale University, New Haven, Conn., 1986. For a description of the survey, see Richard R. Nelson and Richard C. Levin, "The Influence of Science, University Research, and Technical Societies on Industrial R&D and Technical Advance," Research Program on Technological Change, Discussion Paper #3, Yale University, New Haven, Conn., 1986.

in physics and mathematics are picked up and used in fields like electrical engineering and materials science and, through these applied fields, influence industrial technology. Some evidence for our interpretation is provided in Table 2-5. Respondents were asked to rate the relevance of the field itself to their industry. Note that there were many more respondents who rated physics and mathematics as highly important as a field of science than there were respondents who rated university research in these fields as highly important. In our view, this is a crucial distinction that reflects two things. First, the fundamental science learned by industrial scientists and engineers when they attended university plays an important role in industrial R&D problem-solving, even though recent publications in those fields may find little direct use. Second, respondents understand that applicable research findings in fields like electrical engineering and medical science are drawn from and enriched by the more basic sciences.

It is useful to compare these findings with those of two other recent studies that have probed the connection between university research and technical advance in industry. One of these is a series of interviews conducted by the Government-University-Industry-Research Round-table (GUIR Roundtable 1991). The other is a study by Edwin Mansfield (1991, 1–12).

The GUIR Roundtable study was carried out through discussions with seventeen senior industrial research managers who were selected primarily from large, successful industrial companies. Managers were selected from the pharmaceutical and electronics industries, from companies that designed and put together large "systems," and from companies that produced commodities like metals or household products. As one analysis (Gomory 1992) of the interviews indicates, biotechnology is the primary area in which corporate managers look to university research for "inventions." Respondents stated that this was because the technology was new and noted that they believed the direct role played by university research in invention would diminish as the industry matured. We would add that the technology itself was born in a university setting, which is quite unusual. Respondents from electronics companies tended to make a distinction between what they called breakthrough inventions and normal incremental inventions. These respondents took the position that academic research in the field of electronics is often the source of radically new designs and concepts but that the bulk of the total inventive effort and practical payoff came from incremental advances, which were almost exclusively the domain of industrial research, design, problem solving, and development.

Table 2-5. The Relevance of Scientific Fields
to Industrial Technology

Field	Number of Industries Giving Field a High Relevance Score		Selected Industries in Which the Relevance of the Field Is High
	5 or 6	*6*	
Biology	14	8	Drugs, pesticides, meat products, animal feed
Chemistry	74	43	Pesticides, fertilizers, glass, plastics
Geology	4	3	Fertilizers, pottery, nonferrous metals
Mathematics	30	9	Optical instruments, machine tools, motor vehicles
Physics	44	18	Semiconductors, computers, guided missiles
Agricultural science	16	9	Pesticides, animal feed, fertilizers, food products
Applied math/ operations	32	6	Guided missiles, aluminum smelting, research
Computer science	79	35	Guided missiles, semiconductors, motor vehicles
Materials science	99	46	Primary metals, ball bearings, aircraft engines
Medical science	8	5	Asbestos, drugs, surgical/medical instruments
Metallurgy	60	35	Primary metals, aircraft engines, ball bearings

Source: Previously unpublished data from the Yale Survey on Appropriability and Technological Opportunity, Yale University, New Haven, Conn., 1986. For a description of the survey, see Richard R. Nelson and Richard C. Levin, "The Influence of Science, University Research, and Technical Societies on Industrial R&D and Technical Advance," Research Program on Technological Change, Policy Discussion Paper #3, Yale University, New Haven, Conn., 1986.

Respondents discussing drugs that did not emanate from biotechnology stated that university research was almost never the direct source of new products; in virtually all cases the key work was done in industry. However, they also noted that academic research in a number of cases had illuminated the kinds of biochemical reactions that should be sought or had opened the door to a more effective assessment of possible uses for drugs they were testing. These respondents and respondents from several other industries observed that a major function of academic research was to improve the understanding of technologies, particularly new technologies, so that industry could improve them more effectively.

It should be noted that the kinds of local companies that state universities and regional engineering schools traditionally have served were not represented at all, and there was only one executive from a company with products based in agriculture or forestry. That individual did stress the important role of university research to his company.

Mansfield's recent study provides still another window into the role of university research in technical advance in industry. Mansfield asked respondents in seventy-six large American firms what percentage of new products and processes commercialized by their firms from 1975 to 1985 could not have been developed without substantial delay without recent academic research. He then asked for the percentage substantially aided by recent academic research. Table 2-6 summarizes his findings.

Executives in the pharmaceutical industry reported strong dependence on academic research. They stated that over 25 percent of the new drugs commercialized by their companies could not have been developed, or could have been developed only with substantial delay, without academic research. Academic research also substantially aided the development of close to another 20 percent (Mansfield 1991, 1–12). It seems highly likely that pharmaceutical executives interviewed by the GUIR project have accurately characterized the nature of that dependence: Academic research creates knowledge that enables drug companies to search for and develop new drugs more expeditiously but is seldom directly involved in development.

After pharmaceuticals, the reported fraction of new products that are heavily dependent upon academic research for their introduction drops off dramatically. Executives from companies producing information processing equipment and executives from companies producing instruments, report a 10 to 15 percent figure. In the information processing field, it seems likely that most university research contributions

Table 2-6. Percentage of New Products and Processes Based on Recent Academic Research, Seven Industries, United States, 1975–1985

Industry	Percentage That Could Not Have Been Developed Without Substantial Delay in the Absence of Recent Academic Research		Percentage That Was Developed with Substantial Aid from Recent Academic Research	
	Products	*Processes*	*Products*	*Processes*
Information processing	11	11	17	16
Electronics	6	3	3	4
Chemical	4	2	4	4
Instruments	16	2	5	1
Pharmaceuticals	27	29	17	8
Metals	13	12	9	9
Petroleum	1	1	1	1
AVERAGE	11	9	8	6

Source: Data from Edwin Mansfield, "Academic Research and Industrial Innovation," *Research Policy* 20 (1991): 1–12.

are in the form of the prototype radical breakthroughs discussed by GUIR respondents. In the instrumentation field, it seems likely that university contributions come from new or improved instrumentation created by university scientists for their own research uses. The respondents from the metals industry also reported that more than 10 percent of their new products and processes could not have been developed without recent academic research.

A striking finding, which Mansfield does not stress, is that three industries—electrical equipment, chemical products, and oil products—report that only a small percentage of new products (6 percent or less) were significantly dependent upon recent academic research. This does not mean that technical advance in these fields does not rely on science. It does, however, seem to indicate that the science used is not particularly new or presently undergoing academic investigation.

Although the coverage and methodology of the Yale, GUIR, and Mansfield studies are different, these studies do provide a coherent picture of the role of academic research in technical advance in industry. The old service role to local industry, particularly industry tied to agriculture and forest products, is clearly much smaller than it was before World War II, and these industries themselves have dwindled in importance. Even so, the evidence indicates that they continue to depend on academic research for technical advance. The massive funding of research by DOD and kindred agencies shows up clearly in various measures of the contributions of university research to technical advance in electronics. It is also possible to see funding by the NIH in measures of the contributions of academic research in health-related fields. In these fields, however, the development portion of R&D is done by industry. While these studies show that academic research is important to a number of industries, they also show that there are a large number of industries that seem to be relatively untouched by this research. These include basic industries such as steel, autos, and textiles.

WHAT CAN INDUSTRY EXPECT OF UNIVERSITIES AND VICE VERSA?

We began this chapter by noting the significant increase over the past two decades in the fraction of academic research funded by industry and the rapid growth in both the number and the size of university-industry research centers. Many in universities clearly see this as just the beginning, and many of those who are concerned about government policies toward universities anticipate that increased industry funding will reduce the need for government funds. While at first this may sound like a harmony of consistent anticipations and expectations, there are strong reasons to suspect the harmony will fail.

In the first place, many of the academics hoping for further increases in industrial funding also hope this will occur without much change in what they do or how their research is oriented. Many have a firm belief in what has been called the "linear model" of technological advance. They see unfettered research as the basis for technological innovations in industry. While the new government programs buy into some of this, they are insisting more and more on significant industry involvement in the final allocation processes, which means that industry will certainly influence the composition and nature of academic research. The need to assure technology transfer will further link industry and academic research.

Other academics clearly welcome the notion of close ties to industry. Indeed, among some there seems to be a belief that, with sufficient financial support from industry, they can offer a cornucopia of new products and process prototypes and restore the competitiveness of American industry. In their view, American industry needs to become more receptive to the new technology coming out of academia. It is the technology transfer process, according to them, that requires improvement. However, the industry views drawn forth by the GUIR Roundtable suggest considerable skepticism over the ability of academics to contribute directly to industrial innovation, which probably reflects a retreat from the hopeful but less realistic beliefs held in the 1980s. Overall, industry seems to feel that academics should stick to basic research, heed their training functions, and stop thinking of themselves as sources of technology. Such a view suggests that it is highly unlikely industry funding will increase much in the coming years.

We believe that both expectations of what university research, if suitably reoriented, can contribute directly to industrial innovation and expectation of substantial increases in industry funding are equally unrealistic. At the same time, we disagree with those academics and others who argue for a continuation of the status quo. It is time to reconsider what we should expect of the university research system and how the research done under that system should relate to industry. We believe the issue of competitiveness is a serious one, and that the retreat from industrial research in a number of American industries calls for thinking about ways in which American universities can step into the breech.

The end of the cold war has eroded the national defense rationale that has justified government support of university research in a number of fields. However, these fields, which include electrical engineering, computer science, and materials science, are of vital importance to American industry, and industry itself regards academic research in them as a major contributor to industrial innovation. The first order of business, in one view, is to assure that government support of university research in the engineering and applied sciences disciplines is not orphaned by cutbacks in military R&D. It must be clearly understood that government funding in these fields assists American industry. This requires more than a simple change in rhetoric. We need to establish university research support programs that have this express objective and have allocation processes that achieve a sensible allocation of funds, given this objective. This requires advisory committees that are knowledgeable of industry needs and proposal evaluation systems that are sensitive to those needs. Once allocation processes are in line with this objective, discussion can proceed about levels of funding in these fields, including the possibility of increasing or diminishing this funding in certain fields.

As the last twenty-five years of experience with industrial research clearly indicates, if such research is to be fruitful, there must be interaction between those who do research and those who are responsible for product and process design and development. If university research is to assume more of the role of industrial research, there needs to be close links between university researchers and their scientific and technical colleagues in industry. These links already exist to some degree in technologies relating to defense, agriculture, and health. If university research is to play a more helpful role in industrial innovation, such connections need to be further extended and strengthened, which the new university-industry research centers may accomplish.

Based on the surveys and interviews previously discussed, university research in these fields is servicing only a limited range of industries—specifically, those connected with electronics, chemical products, health, and agriculture. Although this is not surprising as these are the fields that government agencies have been supporting for a long time, a policy of broadening the range of industries under which there is a university research base would be quite reasonable. However, if that is to be a policy, it must be a patient policy that looks to practical returns in the long run, not the short.

Does this mean, as some argue, that universities should become more involved in the business of helping industry to develop specific new products and processes? For a number of reasons, we do not think so. As we have noted, the development of the applied sciences and engineering disciplines has led to the establishment of a fruitful division of labor between universities and industry in many technological fields. Universities have taken the responsibility for training young professionals, most of whom go on to work in industry. Academics in these disciplines have performed much of the research that has led to the theories, concepts, methods, and data that are useful to industry in the development of new products and processes. While in some fields this has involved pilot versions of radically new products and processes, as well as research into fundamental scientific questions, by and large it has not involved commercial judgments.

For the most part, university researchers are more poorly positioned than those in industry to see specific operating problems or needs clearly or to be able to judge what may or may not be an acceptable solution to an industrial problem. Indeed, university researchers are almost always insufficiently versed in the particularities of specific product markets to make good decisions about appropriate tradeoffs. Furthermore, such work provides few results that are respected or rewarded in academic circles, unlike research that pushes forward conceptual knowledge in an applied science or engineering discipline. On the other hand, while corporate research laboratories like Bell Labs, IBM Yorktown, DuPont Central Lab, and others have undertaken this type of research and have performed at or above the level of the top universities, the returns from such work are hard to make proprietary. Thus, even if corporate research should experience a total recovery from its recent slump, universities will remain the dominant sites of this research in many fields.

What of the practical problem solving that marked earlier days of American university research, that is, the research on boilers or the processing of ores that used to be quite common on university campuses? This kind of work is still there and often associated with education programs for engineers who will enter industry or with business "incubator" programs at places such as Georgia Tech. It exists on a larger scale and in a more systematic form in institutions that are affiliated with universities but not an integral part of them (e.g., Carnegie-Mellon's Center for Iron and Steelmaking Research or the Forest Products Laboratory at the University of Wisconsin). By and large, these programs have grown up in fields in which industrial research is not strong. They are substitutes for industrial R&D or represent loci for it outside of industry itself. The industries are often, although not always, made up of small firms without R&D facilities and often involve technologies that lack a firm underlying scientific base. As our earlier discussion indicated, university involvement in this kind of research often has its historical origins, and much of its current basis, in training programs. Larger scale research organizations, such as the agricultural experiment stations affiliated with many universities, tend not to be integral parts of the university, but partially detached. Often many of the researchers are not university faculty members, although some may teach courses. On the other hand, their interactions with their industrial clients may be very close. While these kinds of programs can be very valuable to industries and they are an important part of the activities of many universities, after a certain size is surpassed, their locus at universities becomes more a matter of historical happenstance than a source of strength. They could exist just as well as separate organizations.

In any case, we do not think that the emphasis of university research ought to be here or that a policy of federal support of university research that emphasizes contributions to industrial technical advance should be oriented to this kind of work. It is in research, not commercial design and development, that universities excel. While many of the problems of American industry may reside in product and process development and improvement, this is the kind of work industries have to do themselves or have done in specialized industry-linked institutions, which may or may not be associated with universities.

A shift in the emphasis of university research toward the needs of civilian industry can benefit both industry and the universities if it is done in the right way. And that way, in our view, is to respect the division of labor between universities and industry that has grown up with the development of the engineering and applied sciences disciplines. There is no reason to believe that universities will function well in an environment in which decisions need to be made with respect to commercial criteria and every reason to believe that such an environment will damage the legitimate functions of universities, especially if this type of research is carried on as a replacement for, rather than as an addition to, traditional types of research.

References

Cohen, Wesley, Richard Florida, and Richard Goe. 1994. *University Industry Research Centers in the United States*. Pittsburgh: Carnegie Mellon University.

Davis, E. W. 1964. *Pioneering with Taconite*. St. Paul: Minnesota Historical Society.

Gomory, Ralph. 1992. "The Technology-Product Relationship: Early and Late Stages." In *Technology and the Wealth of Nations*, edited by Nathan Rosenberg, Ralph Landau, and David Mowery. Stanford: Stanford University Press.

Government-University-Industry-Research Roundtable. 1991. "Industrial Perspectives on Innovation and Interactions with Universities" (February). Washington, D.C.: National Academy Press.

Levine, David O. 1986. *The American College and the Culture of Aspiration, 1915–1940*. Ithaca: Cornell University Press.

Mansfield, Edwin. 1991. "Academic Research and Industrial Innovation." *Research Policy* 20: 1–12.

National Science Foundation. National Science Board. 1991. *Science and Engineering Indicators–1991.* Washington, D.C.: U.S. Government Printing Office.

Rosenberg, Nathan. 1982. *Inside the Black Box: Technology and Economics.* New York: Cambridge University Press.

Shapley, Deborah and Rustum Roy. 1985. *Lost at the Frontier.* Philadelphia: ISI Press.

Simon, Herbert. 1969. *The Sciences of the Artificial.* Cambridge: MIT Press.

Tocqueville, Alexis de. 1876. *Democracy in America.* Vol. 2. Boston: John Allyn.

von Hippel, Eric. 1988. *The Sources of Innovation.* New York: Oxford University Press.

3

STRATEGIC ALLIANCES AND INDUSTRIAL RESEARCH

David C. Mowery and David J. Teece

FOR THE PAST FIFTEEN YEARS, U.S. industrial research has been in the throes of a restructuring that has changed the position of industrially funded in-house research within the corporate innovation process. A number of central corporate research laboratories have undergone significant cutbacks or, in a few instances, have been eliminated entirely. Since 1980, as Nelson and Rosenberg's chapter in this book and numerous other studies have noted, U.S. firms have expanded their funding for and relationships with university-based research. In addition, numerous domestic and international intercorporate alliances that span R&D, manufacturing, and marketing have sprung up. This chapter considers the motives for and some implications of these trends, which reflect efforts by many U.S. firms to "externalize" a larger share of the industrially financed R&D that formerly was performed within their boundaries.

Some analyses (for example, the National Science Foundation's 1992 study) have speculated that the recent expansion in external research relationships has reduced growth in spending on in-house research and is responsible, at least in part, for declines in the rate of growth in industry-funded R&D expenditures. Although the growth in alliances and research consortia certainly has affected the role of in-house research, we believe that those who think that alliances and consortia can fulfill all of the functions of in-house R&D are mistaken.

The growth in strategic alliances in R&D is part of a broad restructuring of the U.S. national R&D system that involves change in the funding and functions of industry, universities, and government agencies. United States firms were among the pioneers in the development of in-house industrial research laboratories in the late nineteenth and early twentieth centuries. For decades, industrial research laboratories stood at the "heart of the system" of public and private institutions that financed and managed the creation, commercialization, and adoption of new technologies within the U.S. economy (Nelson 1991). Industry accounted for roughly 50 percent of the national R&D investment during much of the postwar period, but it was responsible for performing more than 70 percent of the nation's R&D (National Science Foundation 1992). Thus, today's growth in strategic alliances should be seen in its broader context of declining rates of growth in U.S. industry-financed R&D.[1]

Industrial research laboratories were first established within many large corporations as part of an effort to strengthen central, strategic functions within the firm; that is, to prevent long-range planning and investment decisions from being dominated by day-to-day operating concerns (Chandler 1962; 1977; 1990; see also Teece 1977; 1988). To be successful, however, R&D, like other operations of central corporate management, has to be integrated effectively with both day-to-day and long-range decision making. Many of the problems that have contributed to recent managerial skepticism of corporate research laboratories stem from the failure to integrate R&D strategies with corporate strategies in today's environment in which the demands for rapid response are far more compelling than they were in the beginning of the century.

By itself, outsourcing R&D does not address this failure. Instead, corporate strategists need to manage external and internal R&D activities as complementary activities within a coherent research program that links R&D and corporate strategy. In-house R&D can monitor, absorb, and exploit the results of research performed in research consortia and at other external sites, including universities. Corporate managers also must improve their management of technology transfer and absorption from joint development projects with erstwhile competitors or suppliers. Better management of these relationships can raise the returns to R&D investments.

In contrast to the arguments that portray knowledge transfer and exploitation as virtually costless (Teece 1988; Mowery 1983; Mowery and Rosenberg 1989; Cohen and Levinthal 1990), these arguments are built on a portrayal that emphasizes the costs and importance of managing the transfer and exploitation of scientific and technological information. Finally, as Teece, Pisano, and Shuen (1992) and Prahalad and Hamel (1990) have noted, there is a need for a dynamic view of the firm and the competencies or capabilities for the enduring competitiveness of firms. Although innovation is prominent among the sources of dynamic competitive advantage, the integration of R&D and firm strategy requires an understanding of the role of technology and a discriminating commitment to its support by senior corporate management.

EXTERNAL SOURCING OF R&D

In several respects, today's efforts by R&D managers to expand their links to external sources of new technology have revived an important function of corporate R&D laboratories during much of the period preceding 1940. Early research facilities of such firms as DuPont, Eastman Kodak, General Electric, and AT&T were expected to monitor technological advances occurring elsewhere within their industries and to advise senior management on the acquisition of technologies from other firms and independent inventors (Hounshell and Smith 1988; Reich 1985; Mueller 1962; and Jenkins 1975). In many cases, in-house R&D laboratories modified and commercialized patents or technologies acquired from external sources. In addition, as Nelson and Rosenberg's chapter has pointed out, a number of pre–World War II corporate research laboratories worked with researchers at U.S. universities.

After World War II, however, the outward orientation of many large corporate research laboratories changed. Several factors influenced this shift in R&D strategy, which has been discussed most thoroughly by Hounshell and Smith (1988) for the DuPont Corporation. The wartime demonstration of the power of organized engineering and innovation and the postwar surge in federal R&D contracts led many large firms to develop or expand central research facilities that had weak links with operating divisions. Having been encouraged or requested to do so by federal funders of classified R&D projects, some corporations created autonomous central research "campuses." University researchers also benefited from the expansion in federal research funding and in doing so shifted their attention and fund-raising efforts away from industry (Swann 1988). The tough antitrust policy that emerged in the late 1930s under Thurman Arnold and remained in place through much of the 1970s also made a number of large firms reluctant to seek external sources for new technologies, which had been a key element of their R&D strategies.

Thus, the "golden years" of corporate research described in the introduction to this book were associated with the inward orientation of industrial R&D. While the growth of central corporate research may be associated with improvements in the basic research capabilities of many corporations, all too often, as Hounshell and Smith (1988), Graham (1986), and anecdotal histories of such facilities as the Xerox Corporation's Palo Alto Research Center (Uttal 1983) suggest, this research was not linked effectively to a corporate strategy for its exploitation. In our view, any reorganization or "externalization" of corporate research that does not include restructuring the relationship between corporate strategy and the firm's internal and external R&D investments will not improve innovative performance.

During the 1970s and 1980s, a series of events contributed to the decision of many firms to seek alternatives to exclusive reliance on in-house sources of expertise in the innovation and technology commercialization processes. First, the U.S. antitrust climate changed significantly during the 1980s, as illustrated by the National Cooperative Research Act of 1984 and the 1993 relaxation of federal antitrust restrictions on joint production ventures. Second, the costs of R&D, especially those associated with the development of new products, grew dramatically (in commercial aircraft, for example, new product development costs grew at an average annual rate of more than 10 percent throughout the postwar period), and became increasingly difficult for individual firms to shoulder in an economic environment characterized by high capital costs and intense competition from other domestic and foreign firms. Third, the recovery from the global political and economic upheaval that occurred between 1914 and 1945 meant that the capability to develop and commercialize new technologies had diffused throughout the world by the 1970s (Nelson 1991). Many of the U.S. firms that had dominated R&D and innovation in their industries during the 1950s and 1960s now faced more technologically sophisticated competitors, which increased the financial penalties associated with slow commercialization. Widespread distribution of the technological and nontechnological (marketing and manufacturing) assets needed to bring a new product to market meant that low-cost access to these complementary assets could be achieved most effectively through collaboration with other firms. Finally, scientific and technological advances increased the demands on firms to develop expertise in a wide array of technologies. Firms in food processing and pharmaceuticals, for example, confronted the challenges of biotechnology; telecommunications and computer technologies virtually merged; and advanced materials increased their importance in a broad range of manufacturing industries. Even the largest U.S. corporations, many of which also faced demands for improved financial performance, could not shoulder the costs of in-house development of expertise in an expanding array of strategic technologies. Thus, other firms, consortia, or universities offered possibilities for sharing the costs of developing the required new capabilities.

These forces and others that are specific to each type of collaborative activity have influenced the development of three broad forms of R&D collaboration during the past 15 years. These forms can be characterized as international strategic alliances, precommercial research consortia, and university-industry research collaborations. Collaborative ventures between U.S. and foreign firms now focus on activities, such as joint product development, that did not figure prominently in many of the international joint ventures of the 1950s and 1960s. Domestic consortia of U.S. firms, such as the Microelectronics and Computer Technology Corporation (MCC) have been organized during the past decade to carry out "precommercial" research. University-industry research collaborations now involve larger flows of funds and more U.S. firms.

Each of these forms of collaboration differs somewhat in activities, strategy, and goals. Therefore, the effects of each on in-house corporate research usually differ. International strategic alliances focus mainly on development, production, and marketing rather than precommercial research.[2] Thus far, most domestic collaborations among U.S. firms have concerned research that is not closely linked to specific commercial products. Despite the aspirations of several of them at the time of their founding, these collaborations have rarely focused on basic research. University-industry research collaborations appear to incorporate scientific and engineering research that extends downstream from basic research but typically is not specific to a single commercial product. Thus, two of the three forms of research collaboration do not concern development, the D of R&D that accounts for more than two-thirds of all U.S. R&D spending. In other words, a considerable portion of the research collaboration occurring in U.S. firms involves a relatively small share of their R&D investment.

International Strategic Alliances

International joint ventures have long been common in extractive industries such as mining and petroleum production (Stuckey 1983) and have accounted for a significant share of the foreign investment by U.S. manufacturing firms since World War II.[3] Recently, however, the number of collaborations (between U.S. and foreign enterprises) has grown.

Furthermore, such collaborations now appear in a wide range of industries.[4] The activities that are central to many of these recent ventures, including research, product development, and production for world markets, were absent from most of the ventures of the pre-1975 era, which focused primarily on production and marketing for the domestic market of the non-U.S. firm.

While these ventures are primarily responses to the rising costs and risks of unassisted product development, the growth of technological strengths within foreign firms, the prominent role of nontariff trade barriers in world markets, and government support for the development of advanced technologies, there are other important reasons for their creation. Many recent domestic and international alliances have been formed in the effort to create "bandwagons" behind a particular technical standard. For example, Matsushita's victory over Sony in the Beta-VHS competition in videocassette recorders, for example, owed much to the firm's success in gaining the allegiance of other Japanese and foreign firms to its VCR architecture (Cusumano et al., 1992; Grindley 1990).[5] On the other hand, U.S. firms have created international strategic alliances to improve their access to foreign markets, especially high-technology markets in which governments are engaged in managing trade flows.[6] The search for foreign capital and technological resources also has motivated U.S. firms to enter international joint ventures in industries ranging from semiconductors to steel.[7]

Firms engaged in international strategic alliances need to maintain a strong intrafirm competence in technologies related to the joint venture, for several reasons. Although the central activities of many international joint ventures usually are focused on the development and/or manufacture of specific products or technologies, they provide many opportunities for all parties to learn from their collaborators. According to some scholars (Prahalad and Hamel 1990), some firms, such as NEC of Japan, have developed technology-based core competences relatively inexpensively through their use of joint ventures as learning opportunities. Thus, focusing solely on the completion of the development project may limit opportunities for learning from joint venture partners. Moreover, because many joint development projects produce intellectual property by-products, it is important for firms to negotiate carefully the provisions governing the valuation, exploitation, and sharing of any revenues associated with these by-products.[8]

Exploiting learning opportunities requires an intrafirm capacity to absorb and apply the fruits of the venture. Joint venture participants should create mechanisms for absorbing technology transferred from their partners. The creation of these mechanisms often requires complementary in-house investments. One such investment is the rotation of research and engineering personnel from the firm through a collaborative project. However, it is not enough for an employee simply to capture knowledge (both codified and tacit) or skills from other firms. The employee must be given opportunities to communicate that knowledge to others within the parent firm. This can be done through parallel development and engineering activities within the parent firm.

Because international joint ventures act as vehicles for technology transfer and skills acquisition, the value of the knowledge or capabilities contributed by any single partner depreciates, *ceteris paribus*. As technology is transferred through a collaborative venture, learning by the other participants reduces the value of the technological capabilities that were originally unique to one or another participant. Depreciation may be even more rapid in ventures in which one firm contributes its marketing knowledge and network or other "country-specific" expertise. Although an alliance may be an essential means to gain access to new knowledge, as the other participants improve their knowledge of the markets in which this partner has specialized, they are likely to choose to continue without it.[9] This fact has played a role in the breakup of a number of collaborative ventures between Japanese and U.S. producers of auto parts. As the Japanese partners in these ventures have gained knowledge about local markets and production conditions (particularly when selling to Japanese transplant operations in the United States), they have withdrawn from the joint venture to continue independently (Phillips 1989). Although technology-based assets are likely to depreciate more slowly, especially if technology transfer is closely managed, Hamel, Doz, and Prahalad (1989) suggest that process technologies are less easily exploited by other participants than product technologies, which are more transparent to venture partners.[10]

Depreciation in the value of assets within a joint venture is no less inevitable than depreciation of physical capital assets within a manufacturing plant. In both cases, participants must take steps to reduce erosion in the value of their contribution and, at the same time, seek ways to offset the consequences of such depreciation. Intrafirm investments in technology development are essential to the creation or maintenance of the quality of the technological assets contributed to the joint venture. Therefore, participant firms must sustain in-house technology development activities in product lines and technologies that are related to the joint venture and managers need to pursue ways to offset the depreciation that will occur, for example, by exploiting learning opportunities. If the collaborative venture aids in the establishment of a product design as a market standard or a venture with an established firm provides an endorsement of a technology, some of the detrimental effects of this depreciation also can be reduced. When a firm provides a static asset like market access, the collaborative venture may function most effectively as a means for exiting the industry or as a temporary channel for learning process and product technologies.

For many U.S. firms, joint ventures involve closer work with suppliers. In these user-supplier ventures, more responsibility for the development of components to meet specific performance parameters is delegated to the supplier and the risks and costs of development are shared. In the semiconductor industry, these ventures often team U.S. and Japanese firms and generate considerable product-specific and technology-specific know-how and intellectual property. Shuen's research (1993) suggests that the failure of some U.S. semiconductor firms to invest sufficient resources in monitoring and absorbing jointly developed intellectual property has reduced their returns from these relationships. It would appear that U.S. managers need to broaden the channels through which they obtain technologies from external sources. Know-how and technological capabilities do not come exclusively from formal "horizontal" joint ventures. They also flow from marketing, supplier, and numerous other relationships. Moreover, as firms come to rely more heavily on these relationships, they need better channels for transfer and absorption.

In addition to improving market access, reducing risk, and lowering the costs and time required for new product development, international joint ventures in product development can raise the efficiency of a firm's internal R&D. These ventures allow firms to exercise greater selectivity in their in-house technology investments. However, selectivity must be based on a careful analysis of the firm's strengths and weaknesses, and the long-run impact of reducing investment in specific technologies on corporate strategy must be clearly laid out. In other words, entry into an international joint venture should be based on an integrated analysis of technological and firm strategy. Such an assessment should include some evaluation of the competitive sensitivity of specific parts—the firm's "crown jewels"—of its technology portfolio. The uncertainties associated with technology-based competition mean that any such analysis is likely to rely more heavily on the construction and evaluation of scenarios than on the illusory precision of quantitative estimates. Moreover, the firm pursuing this selective approach still will require in-house expertise to evaluate the strengths and weaknesses in the technologies of its prospective collaborators.

Precommercial Research Consortia

Research consortia, funded entirely or in part by industry funds and focusing on precommercial research activities, are a recent innovation in U.S. firms. In most cases, these consortia have involved U.S. firms only (some consortia that are funded in part from public sources, for example, SEMATECH, have formal policies excluding foreign firms). More than 450 such ventures have been registered with the U.S. Justice Department through 1994 under the terms of the 1984 National Cooperative Research Act (Evan and Olk 1990; Werner 1992; Link 1995).

Several of the most widely publicized consortia founded during the first half of the 1980s, such as MCC, were established in response to Japanese cooperative research programs (Peck 1986), particularly the VLSI program of the 1970s and the Fifth Generation Computing project that was undertaken during the 1980s as a successor to the VLSI program. Computer industry executives in the United States concluded that Japanese cooperative programs supported the type of long-range research that no single firm would undertake; projects like MCC were created to fill this void in the U.S. computer industry.

The short history of MCC and the experiences of consortia in other industries (for example, the Electric Power Research Institute, which serves the U.S. electric utility industry) suggest that research consortia rarely sustain a long-range focus but rather shift their focus to research on generic technology issues of more immediate interest to member firms. Both SEMATECH and the National Center for Manufacturing Sciences are now pursuing technology-focused research that seeks to improve vertical relationships between users and suppliers of capital equipment (Grindley et al. 1994). Interestingly, the Japanese industry consortia that sparked U.S. concern have rarely undertaken basic research. Instead, they have focused on technology development and dissemination among their members. Japan's Fifth Generation Computing project, which pursued a longer-term research agenda, has been relatively unsuccessful.

Like international strategic alliances, most research consortia focus on technology development. In contrast to alliances, however, agendas of consortia usually are not highly product-specific. Indeed, the example of SEMATECH suggests that agreement among consortium members on an agenda that focuses on specific (and often proprietary) product or process technologies may be difficult if not impossible to achieve (Grindley et al. 1994). For this reason, the near-term competitive risks from participating in a consortium may be less significant than the risks associated with an international joint venture. The financial costs of consortium participation also are likely to be lower than the costs of a joint venture. Therefore, decisions on consortium participation, management, and so forth may raise fewer long-term corporate strategy issues and do not require the same degree of senior management participation as decisions on international joint ventures.

Although there are a number of important differences in their goals and structure, the requirements for maximizing technology-based benefits from consortia closely resemble the requirements associated with strategic alliances. Complementary investments in R&D within member firms, the creation of channels of communication and technology transfer with the consortium, and the development of an in-house receptor are necessary to increase the returns from participation. For example, MCC's reliance on its own research staff, in contrast to SEMATECH's use of assignees from member firms, made it difficult for member firms to absorb the results of MCC research. The complex structure of MCC, which established barriers to some firms' access to certain research areas, further impeded technology transfer.

University-Industry Research Collaboration

Much of the recent expansion in domestic research collaboration involves a renewal of the link among state governments, publicly supported universities, and industry that languished during the post-1945 period (Mowery and Rosenberg 1993). The huge size, decentralized structure, and research-intensive character of the American universities are unique and increase the potential payoff from collaboration between universities and industry. Nevertheless, clearly defined "deliverables" often are of secondary importance in successful university-industry collaborations (Mowery and Rosenberg 1989; Rosenberg and Nelson 1994).

Thus far, university-industry collaborations appear better suited to the support of long-term, precommercial research than interfirm consortia. This tentative conclusion is based on the tendency of consortia to shift their agenda toward near-term research. The personnel flows between firm and research laboratory that often figure prominently in university-industry collaborations also aid communication between university and corporate research establishments.

Because U.S. universities include education as well as research in their activities, they are important sources of scientific and engineering personnel for industry. Firms can use collaborative ventures as filters for hiring research personnel, since the ventures allow them to observe the performance of potential researchers before making employment commitments. Moreover, the hiring of graduates of these programs facilitates the transfer of knowledge and technology even more effectively than does the rotation of industry personnel through university research facilities. Given the interdisciplinary character of current industrial technological and research challenges, the training of research personnel is an especially important benefit to industry that may emerge from industry-university collaborations. Firms in the semiconductor, biotechnology, or robotics industries now need individuals with interdisciplinary research training. Industrial funding, like federal government research support during the 1960s and 1970s, can aid in the establishment of university interdisciplinary research and education programs, which are notoriously difficult to develop without external funds.[11]

Thus, through interdisciplinary research and education programs, collaboration with universities can provide firms with "windows" for monitoring novel research areas and filters for hiring research and technical personnel. While the results of university-industry research collaborations may rarely be applied directly to commercial innovation, by improving access to university research, they can improve the efficiency of in-house research activities. As Nelson and Levin (1986) and David, Mowery, and Steinmueller (1992) have argued, many of the economic benefits of university research and other basic research are realized through the ability of the research findings to improve the efficiency of applied research; that is, basic research results lead to a better informed, and therefore more efficient, "search" process in technology development. It is the general knowledge produced by this research, rather than any specific discoveries, that provides many of the economic benefits.[12]

What does this characterization of the economic payoffs from these collaborations imply for managers who wish to increase the competitive benefits arising from such relationships? As in the case of strategic alliances and research consortia, the creation and maintenance of good channels of communication and transfer are critical and require both the hiring of graduates and the rotation of firm personnel through university research facilities. Managers also must maintain "receptors" within the firm to absorb and apply university research findings to technology development. One of the few empirical studies of the role of external basic research in innovative performance found that pharmaceuticals firms with strong intrafirm "academic," or basic, research capabilities more successfully exploited such research than firms lacking these capabilities (Gambardella 1992). In other words, university-based research must complement in-house research activities. Without some capability to understand and exploit the results produced in collaborative research relationships, the returns to these external investments are likely to be low. University research collaborations may allow for greater selectivity in the in-house basic research agenda, but they cannot be effective without complementary in-house research activities.

CONCLUSION

This chapter has provided a taxonomy for understanding the external research relationships that have recently sprung up within U.S. industrial research and has suggested some ways in which the management of these relationships influences in-house research activities and innovative performance. Although some recent experiments in strategic alliances represent a revival of the earlier, outward-oriented R&D strategy followed by many of the pioneers in U.S. industrial research, the outcomes of these new undertakings remain uncertain. The restructuring of U.S. corporate research is likely to continue for some time because of severe competitive and financial pressures on U.S. firms.

In order to utilize research collaboration as an effective competitive solution, managers must define the problem they are addressing. Undertaking external R&D relationships primarily or solely as a means of reducing the corporate R&D budgets, for example, may do little to improve the long-term returns to corporate R&D investments. The disappointing returns to many R&D investments reflect a frequent failure to maintain links between R&D priorities and corporate strategy rather than excessive levels of R&D investment. This problem is not addressed by external R&D, and without a solution to it, external R&D

may well fail. Research collaboration provides opportunities for R&D cost reduction or improved market access, but entry into such ventures solely for these purposes is ill-advised. Collaborative research ventures should be undertaken and managed for their potential to strengthen the capabilities that underpin firms' competitive strength.

While external research ventures can support learning from other firms or research institutions and a more focused in-house R&D agenda and budget, to do so successfully corporate strategy and R&D priorities must be well integrated. The failure to integrate may then prove to be counterproductive causing R&D collaboration to erode, rather than strengthen, corporate competitive advantage. To improve the payoff from external research ventures, a firm must invest in activities that facilitate the inward transfer of knowledge and technology, but this is insufficient by itself. Complementary investments in intrafirm R&D are also necessary to provide opportunities for the exploitation and "absorption" of the fruits of external research. As noted earlier, managing these external research relationships as complements to an in-house research portfolio can facilitate a more efficient allocation of intrafirm R&D investments among technologies or strategic opportunities.

Successful management of external research relationships also requires a good fit between the type of external research venture and corporate or business unit goals. Using a university-industry collaboration to accelerate the development of a product or using a precommercial research consortium to strengthen basic research capabilities is likely to prove disappointing. Furthermore, the failure to recognize that these activities may yield multiple types of benefits can reduce their payoffs. Joint product development ventures often yield significant learning by-products and may develop intellectual property in related areas or technologies. Capturing these benefits and managing their exploitation requires careful consideration of different approaches to organizing and negotiating the terms of a venture.

As noted earlier, the focus of many of these collaborations on the lower-cost phases of industrial innovation means that their effects on in-house R&D spending may have been overstated. Moreover, as noted repeatedly, the successful exploitation of external R&D requires complementary in-house R&D investments. In light of these conclusions, recent flat trends in industrially funded R&D spending may not reflect improvements in efficiency or productivity as a result of collaborations. Indeed, policies that seek to improve industrial competitiveness by encouraging externally based research networks (for example, tax incentives or direct subsidies) may not offset the effects of declining intrafirm R&D if the impact of these collaborations on innovative performance depends on intrafirm R&D investment. These policies also cannot address performance problems that reflect deficiencies in the integration of corporate strategy and R&D management and priorities.

Notes

1. Real annual growth rates in industry-financed R&D spending have declined since 1984. Rates reached zero in 1986–87 and in 1990–91 (National Science Foundation 1992). The share of GDP accounted for by industry funded R&D in the United States has lagged behind the share of GDP in both Germany and Japan by a widening margin during the past decade.

2. A number of regional programs in Western Europe, such as ESPRIT and EUREKA, focus on precommercial research. For purposes of this discussion, however, these consortia within an economically unified region are treated as similar to domestic research consortia.

3. Indeed, although their relative importance has declined, the absolute size and number of joint ventures in the extractive industries remain substantial and may have increased. Karen Hladik's analysis (1985) of data from the Harvard Multinational Enterprise Project concluded that 39 percent of the number of foreign subsidiaries established by U.S. manufacturing firms from 1951 to 1975 were joint ventures. Benjamin Gomes-Casseres (1988) analyzed these data and found a significant decline in the share of joint ventures within U.S. multinationals' international subsidiaries from 1961 to 1968, followed by a resumption of growth in the share of joint ventures from 1969 to 1975.

4. Kathryn Harrigan (1984) concluded that domestic joint ventures involving U.S. firms had grown during the previous decade. In the 1960s, joint ventures were concentrated in the chemicals, primary metals, paper, and stone, clay, and glass industries, but they now extend beyond these sectors. Karen Hladik (1985) found significant growth from 1975 to 1982 in the number of international joint ventures involving U.S. firms. This trend has almost certainly continued through the present.

5. These alliances are not always successful, however. The Advanced Computer Environment (ACE) failed to establish the RISC microprocessor architecture developed by MIPS Computer Systems as a standard. See Jonathan Khazam and David C. Mowery (1994).

6. U.S. firms are not the only ones to use joint ventures in this manner. Foreign firms have found that joint ventures can improve their access to U.S. markets.

7. Historically, in industries such as telecommunications, technical standards have been established through multilateral or plurilateral negotiations

among governments. These negotiations are heavily affected by governments' perceptions of the effects of a particular standard on the competitive fortunes of "national champions," many of which are government-owned or controlled. By establishing a network of international alliances, U.S. telecommunications firms have sought to gain advantage in negotiations over government-sponsored technical standards. See David C. Mowery (1989).

8. David J. Teece (1992) argues that the importance of such by-products and the development of provisions for their exploitation may favor the use of a shared-equity ownership structure for joint ventures.

9. Michael E. Porter and Mark B. Fuller (1986) have observed that collaborative ventures centered on marketing "may be particularly unstable, however, because they frequently are formed because of the access motive on one or both sides. For example, one partner needs market access while the other needs access to product. As the foreign partner's market knowledge increases, there is less and less need for a local partner" (p. 334).

10. "The type of skill a company contributes is an important factor in how easily its partner can internalize the skills. The potential for transfer is greatest when a partner's contribution is easily transported (in engineering drawings, on computer tapes, or in the heads of a few technical experts); easily interpreted (it can be reduced to commonly understood equations or symbols); and easily absorbed (the skill or competence is independent of any particular cultural context). . . . Western companies face an inherent disadvantage because their skills are generally more vulnerable to transfer. The magnet that attracts so many companies to alliances with Asian competitors is their manufacturing excellence—a competence that is less transferable than most" (Gary Hamel, Yves Doz, and C. K. Prahalad 1989, 136). The converse is also true. A central technological asset contributed by Boeing to its collaborative ventures with Japanese firms is its expertise in production technology and in the management of fluctuations in production volume for commercial airframes (Mowery 1987).

11. The departmental structure of U.S. universities appears to make this task easier than it is in many Western European universities: "Among the factors cited to explain West Germany's slow entry into commercial biotechnology is an educational system that prevents the kind of interdisciplinary cooperation that is viewed by most experts as essential to the development of this field. In particular, the traditional separation of technical faculties from their arts and sciences counterparts means that process technicians, usually located in the technical schools, rarely come into contact with colleagues holding university appointments in biochemistry or microbiology" (U.S. House of Representatives, Office of Technology Assessment 1985, 424).

12. D. M. Hercules and J. W. Enyart (1983, 7) report that the following four areas of collaboration had very high potential payoffs and currently lacked sufficient activity: (1) lectureships by academic scientists at industrial sites; (2) student interns at industrial sites; (3) continuing education programs at

industrial sites; and (4) corporate support for employees to obtain advanced degrees. Note that none of these areas involve significant transfers to industry of intellectual property or other deliverables.

References

Chandler, Alfred D., Jr. 1962. *Strategy and Structure.* Cambridge, Mass.: MIT Press.

Clark, Kim B. 1989. "What Technology Can Do for Strategy." *Harvard Business Review* (November–December): 94–8.

Cohen, Wesley M., and David A. Levinthal. 1990. "Absorptive Capacity: A New Perspective on Learning and Innovation." *Administrative Sciences Quarterly* 35: 128–52.

Cohen, Wesley, Richard Florida, and Richard Goe. 1994. *University-Industry Research Centers in the United States.* Pittsburgh: Carnegie Mellon University.

Cusumano, Michael, Yorgis Mylonadis, and Richard Rosenbloom. 1992. "Strategic Maneuvering and Mass-Market Dynamics: The Triumph of Beta over VHS." *Business History Review* 66: 51–94.

David, Paul A., David C. Mowery, and W. Edward Steinmueller, "Analyzing the Economic Payoffs to Basic Research." *Economics of Innovation and New Technology* 2: 73–90.

Evan, William M., and Paul Olk. 1990. "R&D Consortia: A New Organizational Form." *Sloan Management Review* 31 (Spring): 37–46.

Feller, I. 1990. "Universities as Engines of R&D-Based Economic Growth: They Think They Can." *Research Policy* 19: 335–48.

Gambardella, A. 1992. "Competitive Advantages from In-House Scientific Research: The U.S. Pharmaceutical Industry in the 1980s." *Research Policy* 21: 391–407.

Ghemawat, P., M. E. Porter, and R. A. Rawlinson. 1986. "Patterns of International Coalition Activity." In *Competition in Global Industries* edited by Michael E. Porter. Boston, Mass.: Harvard Business School Press.

Gomes-Casseres, Benjamin. 1988. "Joint Venture Cycles: The Evolution of Ownership Strategies of U.S. MNEs, 1945–75." In *Cooperative Strategies in International Business,* edited by F. J. Contractor and P. Lorange. Lexington, Mass.: Lexington Books.

Graham, Margaret B. W. 1986. *RCA and the Videodisc: The Business of Research.* New York: Cambridge University Press.

Graham, Margaret B. W., and Bettye H. Pruitt. 1990. *R&D for Industry: A Century of Technical Innovation at Alcoa.* New York: Cambridge University Press.

Grindley, P. 1990. "Winning Standards Contests: Using Product Standards in Business Strategy." *Business Strategy Review* 1 (Spring): 71–84.

Grindley, P., David C. Mowery, and B. Silverman. 1994. "Sematech and Collaborative Research: Lessons in the Design of High-Technology Consortia." *Journal of Policy Analysis and Management* 13: 723–58.

Hamel, Gary, Yves Doz, and C. K. Prahalad. 1989. "Collaborate with Your Competitors—and Win." *Harvard Business Review* (January–February): 133–39.

Harrigan, Kathryn R. 1984. "Joint Ventures and Competitive Strategy." Working paper, Graduate School of Business, Columbia University, New York.

Hercules, D. M., and J. W. Enyart. 1983. "Report on the Questionnaire on Current Exchange Programs Between Industries and Universities." Council on Chemical Research, University-Industry Interaction Committee.

Hladik, Karen. 1985. *International Joint Ventures.* Lexington, Mass.: D.C. Heath.

Hounshell, David A., and John Kenly Smith. 1988. *Science and Corporate Strategy: Du Pont R&D, 1902–1980.* New York: Cambridge University Press.

Jenkins, Reese V. 1975. *Images and Enterprise: Technology and the American Photographic Industry, 1839–1925.* Baltimore: The Johns Hopkins University Press.

Khazam, J., and D. C. Mowery. 1994. "The Commercialization of RISC: Strategies for the Creation of Dominant Designs." *Research Policy* 23: 89–102.

Link, A. N. 1995. "Research Joint Ventures: Patterns from *Federal Register* Filings." Economics working paper, Center for Applied Research, Bryan School of Business and Economics, University of North Carolina-Greensboro.

Mowery, David C. 1983. "Economic Theory and Government Technology Policy." *Policy Sciences* 16, no. 2: 27–43.

———. 1987. *Alliance Politics and Economics: Multinational Joint Ventures in Commercial Aircraft.* Cambridge, Mass.: Ballinger.

Mowery, David C., and N. Rosenberg. 1989. *Technology and the Pursuit of Economic Growth.* New York: Cambridge University Press.

———. 1993. "The U.S. National Innovation System." In *National Innovation Systems: A Comparative Analysis,* edited by Richard R. Nelson. New York: Oxford University Press.

Mueller, W. F. 1962. "The Origins of the Basic Inventions Underlying Du Pont's Major Product and Process Inventions, 1920 to 1950." In *The Rate and Direction of Inventive Activity,* edited by Richard R. Nelson. Princeton, N.J.: Princeton University Press.

National Science Foundation. National Science Board. 1992. Committee on Industrial Support for R&D. *The Competitive Strength of U.S. Industrial*

Science and Technology: Strategic Issues. Washington, D.C.: U.S. Government Printing Office.

Nelson, Richard R. 1990. "U.S. Technological Leadership: Where Did It Come From and Where Did It Go?" *Research Policy* 19: 117–132.

———. 1991. "Capitalism as an Engine of Progress." *Research Policy* 20: 193–214.

Nelson, Richard R., and Richard C. Levin. 1986. "The Influence of Science University Research and Technical Societies on Industrial R&D and Technical Advance," Research Program on Technological Change Policy discussion paper 3, Yale University, New Haven, Conn.

New York Times. 1992. "University of California Proposes Laboratory-to-Marketplace Link," 11 December, A14.

Pavitt, Keith. 1991. "What Makes Basic Research Economically Useful?" *Research Policy* 20: 109–19.

Peck, Merton J. 1986. "Joint R&D: The Case of the Microelectronics and Computer Technology Corporation." *Research Policy* 15: 219–32.

Phillips, S. 1989. "When U.S. Joint Ventures with Japan Go Sour." *Business Week* (24 July): 30–31.

Porter, Michael E., and Mark B. Fuller. 1986. "Coalitions and Global Strategy." In *Competition in Global Industries*, edited by M. E. Porter, Boston: Harvard Business School Press.

Prahalad, C. K., and Gary Hamel. 1990. "The Core Competence of the Corporation." "Harvard Business Review" (May–June): 79–91.

Reich, Leonard S. 1985. *The Making of Industrial Research: Science and Business at GE and Bell, 1876–1926*. New York: Cambridge University Press.

Rosenberg, Nathan. 1990. "Why Do Firms Do Basic Research (With Their Own Money)?" *Research Policy* 19: 165–74.

Rosenberg, Nathan, and Richard R. Nelson. 1994. "American Universities and Technical Advance in Industry." *Research Policy* 23: 323–48.

Servos, John W. 1994. "Changing Partners: The Mellon Institute, Private Industry, and the Federal Patron." *Technology and Culture* 35 (April): 221–57.

Shuen, Amy S. 1993. "Co-Developed Know-how Assets in Technology Partnerships." Haas School of Business, University of California, Berkeley.

Stuckey, J. S. 1983. *Vertical Integration and Joint Ventures in the Aluminum Industry*. Cambridge, Mass.: Harvard University Press.

Swann, John P. 1988. *Academic Scientists and the Pharmaceutical Industry: Cooperative Research in Twentieth-Century America*. Baltimore: The Johns Hopkins University Press.

Teece, David J. 1977. *The Multinational Corporation and the Costs of International Technology Transfer*. Cambridge, Mass.: Ballinger.

————. 1988. "Technological Change and the Nature of the Firm." In *Technical Change and Economic Theory,* edited by G. Dosi, C. Freeman, R. Nelson, G. Silverberg, and L. Soete. London: Frances Pinter.

————. 1992. "Competition, Cooperation, and Innovation: Organizational Arrangements for Regimes of Rapid Technological Progress," *Journal of Economic Behavior and Organization* 18, no. 1: 1–25.

Teece, David J., Gary Pisano, and Amy Shuen. 1992. "Dynamic Capabilities and Strategic Management." Working paper, Haas School of Business, University of California, Berkeley.

U.S. House of Representatives. 1985. Office of Technology Assessment. *Commercial Biotechnology: An International Analysis* (Washington, D.C.: U.S. Government Printing Office, 1985).

Uttal, Bro. 1983. "The Lab that Ran Away from Xerox." *Fortune* 108 (5 September): 97–102.

Wall Street Journal. 1991. "U.S.'s DNA Patent Moves Upset Industry," 22 October, B4.

Werner, Jerry. 1992. "Technology Transfer in Consortia." *Research-Technology Management* 35, no. 3: 38–43.

PART TWO

PERSPECTIVES FROM PRACTICE

4

RESEARCH AND CHANGE MANAGEMENT IN XEROX

Mark B. Myers

CORPORATIONS HAVE ENTERED an era in which success will be determined largely by their ability to manage change brought about by the dynamic synergy between new uses and rapid increases in the power of technology and reductions in the costs of that technology. The ability to couple technology and markets under rapidly changing and uncertain conditions will be the key to future business prosperity. In technically sophisticated markets, the ability to participate in these opportunities will hinge on the research organizations of corporations. Xerox Corporation's past has been, and its future will be, shaped by its ability to respond to change. This chapter examines the role of Xerox's research organization in helping the corporation to navigate the seas of change and how that role is changing as the organization seeks greater influence over the company's course.

The Xerox story is about change. Over the corporation's short history, which dates from about 1960, the swings of its business experience have been the focus of an extraordinary amount of analysis and writing. John Desseaur's *My Years at Xerox: The Billions Nobody Wanted* (1971) chronicles the identification and development of the xerographic technology that revolutionized office work and precipitated Xerox's explosive growth. Smith and Alexander's *Fumbling the Future* (1988) relates the story of Xerox Palo Alto Research Center's (PARC) invention of the personal computing paradigm that spurred another revolution in the workplace from which Xerox derived almost no economic benefit. Jacobson and Hillkirk's *Xerox American Samurai* (1986) and Kearns and Nadler's *Prophets in Dark* (1992) describe the corporation's return to competitive position. These are stories of growth, failure, and recovery.

The growth was extraordinary. Between 1960 and 1993, Xerox grew from a small, regionally based imaging company in Rochester, New York, to a multinational corporation with a market presence in every major economic region of the world. In this time, revenues increased from $37 million to $15 billion. A glance at Xerox's revenue performance by decade, however, is revealing. During the 1960s, revenue grew at a compound rate of 44 percent; during the 1970s, the rate was 16 percent; and during the 1980s, the rate was 3 percent. After a decade and a half of growth, all indicators of corporate performance were moving in the wrong direction. The 1980s brought an onslaught of tough competition across all segments of Xerox's main copier businesses, and the corporation's substantial investments in the emerging computer workstation industry were not successful. Clearly, the corporation was deeply troubled. Indeed, it was so troubled that David Kearns, the CEO at the time, did not feel it could be successful solely in the office products business in the face of mounting competition and diversified the corporation into financial services.

Throughout the 1980s, Kearns guided Xerox through a comprehensive effort in total quality management (TQM), employing all of TQM's tools across every aspect of the corporation's business. As a result, the erosion in market share was halted, significant new reprographic and printing products were introduced, and key financial performance measures were significantly improved (Kearns and Nadler 1992). These moves and others reestablished confidence in the corporation's ability to be a competitive leader in reprographic businesses.

Two important strategic positioning studies were undertaken in 1985 and 1990. The first, led by Kearns, defined the actions required to further improve the corporation's financial performance by assuring success in core reprographic businesses and exiting businesses that were not succeeding. It was this study that led to Xerox's extraordinarily difficult decision to exit the manufacture and marketing of its proprietary workstation and network server hardware systems. Xerox PARC had created the radical discontinuity of client-server computing only a few years earlier, and it appeared that the corporation was retreating from the digital transformation of the workplace.

The second study, led by Paul Allaire, the new CEO, and his senior management team, defined the business that Xerox aspired to be in the year 2000. The copier, or "marks-on-paper," company recast itself as "the document company," a supplier of productivity-boosting products and services. This vision of Xerox as a document company, rather than a copier company, was founded on its core strengths in image-intensive reprographics and digital printing and an aspiration to offer integrated document services. It anticipated and built upon a major transition from analog to digital technology as the basis for document-intensive work processes. Xerox's Docutech digital reprographic system, digital color copiers, and networked printer products are the initial fruits of the corporation's revised focus.

THE HISTORY OF XEROX RESEARCH

The development of Xerox's research organization and the corporation's good and bad times were concurrent. By most measures, however, the corporation's investment in research created one of the most innovative industrial research organizations in the world. The roots of the present research organization were first nourished by John Bardeen, a member of the board of directors. Bardeen urged the creation of a central research laboratory, which led, around 1962, to the recruitment of a number of senior research managers from General Electric's Schenectady Research Center. The foundation of the research organization was laid in Webster, New York, in 1964 when John Desseaur, then vice president of R&D, dedicated the Webster Research Center (WRC), the corporation's first building devoted to a central research function. The size and reputation of the corporation's research organization grew rapidly between 1968 and 1986. Under the leadership of Jack Goldman and George Pake, research investment in Webster was maintained and new research centers were established in Palo Alto (1970) and in Canada (1974). Under William Spencer, and later, this author, research laboratories were established in Rank Xerox in Europe and research was initiated by Fuji-Xerox in Japan.

The growth of the research organization under Goldman and Pake transcended two eras of corporate research philosophy. While that growth was along the lines of the growth found in such great centralized research organizations as GE, AT&T, IBM, and DuPont, the organization was inherently more decentralized as geographically distributed centers embraced the missions, values, and cultures that reflected their particular competencies (for example, imaging sciences in Rochester, digital systems in Palo Alto, and chemistry in Canada). Together, the centers formed a world-renowned research organization. Highly visible contributions in information technology from PARC gave rise to a new computing paradigm that became populated by an amazing array of techniques and technologies, including personal computing, client-server architecture, graphical user interfaces, local area networks, laser printing, bit maps, page description languages, and object-oriented programming. At all three research centers work in the physical sciences was highly visible to the external scientific community. From 1981 to 1991, the Xerox research organization ranked among the ten most influential academic and industrial research institutions in the United States as measured by frequency of reference to the Xerox papers in research publications (Garfield 1993, 8).

Research laboratories were operationally and organizationally separate from the corporation's large marketing, product development, and manufacturing organizations. The laboratories had the support of and direct access to senior management. Conversations between the head of research and Xerox's chairman or vice-chairman to whom the head of research reported, established the level of research investment without the investment trade-off pressures of the corporate budgeting processes. This access enabled research investment to increase at a compound rate of 8 percent annually between 1976 and 1986, when financial performance indicators were in marked decline.

A deeply embedded management philosophy was responsible for the extraordinary technical successes of the laboratories (Pake 1986). Among the principal tenets of this philosophy were:

1. Recruit the best, most creative researchers you can find.

2. Give the researchers the most supportive environment you can provide, including ample amounts of the most advanced instrumentation.

3. Work the business needs of the corporation into the program through selective budgetary preferences.

To these, Pake added "Make a conscious effort to imitate the best research in universities by providing an intellectual environment" and observed, "selecting the research to work on and allocating the resources to do it is the essence of what we call research management. It draws upon the combination of technical knowledge, business strategies, research experience, understanding of the psychological makeup of research scientists, and, above all, technical tastes" (Pake 1986). In the eyes of Pake and his predecessors, management's role was to communicate broad overall goals and allow talented and creative professionals to fill in the details of the research agenda. Ample support to build real systems to test their ideas and minimum corporate constraint on the freedom to formulate questions encouraged Xerox's research community to think about problems in unconventional ways. The results spoke for themselves. Out of this leadership philosophy and generous funding emerged a prominent research organization.

The chairman's confidence in and support of research were not shared by senior executives of the corporation's functional organizations; their business and personal relationships with research management were often contentious and competitive. Indeed, the chief executive office continuously faced investment decisions that involved deep divisions between the research community's expert opinions and the opinions held by those responsible for the market or product development processes.

Research management recalled as watershed events instances in which it dramatically intervened at the last moment to affect the direction of a decision (Perry and Wallich 1986, 67). One notable intervention by Jack Goldman involved senior management in a reversal of a digital printing technology decision that had favored a CRT input over the modulated laser-addressing technology supported by research. These instances became part of the lore of research management and a model of the role it should serve. While history has justified specific instances of intervention, the role cast the research organization as an outsider given to frequent attacks on the decisions of organizations that were responsible for product delivery or market development. Resulting tensions periodically gave rise to proposals to limit research's independence by downsizing investment in its activities or by splitting the organization's centers and establishing reporting relationships to the functional organizations.

By the early 1980s, Xerox's research organization had become noted for its innovative new concepts that had significant impact in the marketplace but earned the corporation little economic benefit. The organization's growing reputation for innovation became a source of both pride and tension for research management as it appeared to some that the reputation had little relationship to the corporation's general performance (Uttal 1985).

Rethinking the Role of Xerox Research

The relationship of research to the management of change at Xerox warrants a closer examination. Three research themes connect the major challenges the corporation faced during its history of change:

1. Aggressively establish and grow the company by exploiting the xerographic copying technology in the office.

2. Seek technical performance advantages in marking technologies that counter the copying market's erosion to competition.

3. Create a new technology advantage beyond xerography that offers growth opportunities in areas lacking competition.

During the extraordinary growth of Xerox's reprographic business, the research organization contributed enormously to moving the corporation's core xerographic technology from a speculative invention base to a predictive science base. This permitted the creation of generations of increasingly sophisticated high-performance light lens and digital reprographic, printing, and color products that could not have been designed on the basis of qualitative information alone. Indeed, over the past decade and a half, work on laser printing and digital reprographics has supported Xerox's creation of new multibillion dollar markets.

Xerox's exploitation of its xerographic technology was initially supported by a large and growing patent portfolio that protected it from competition. The development of this intellectual property base was a major early focus of the Webster Research Center. However, patent protection ended in 1973 with the Federal Trade Commission's consent decree, which required Xerox to cross-license all existing and future xerographic patents to all competitors for a period of ten years. For the first time, Xerox faced competition. The initial competition was from IBM and Eastman Kodak. These respected, technically sophisticated, well-resourced competitors created copier and duplicator products similar to Xerox's, but they also approached the marketplace in the same manner as Xerox. It was, then, Japanese, not domestic, competition that occasioned Xerox's dramatic loss of market share in its core copier business. With fewer technical resources than Xerox and its domestic competitors and little market recognition, Japanese firms creatively redefined the market to low-end desktop copying machines based on high-quality, low-cost design and manufacturing competencies. These competencies were unanticipated by Xerox and most U.S.-based corporations. Xerox's research organization had initiated activities in VLSI design and precision mechanical design for laser scanning and rotating memory disks but had chosen not to pursue methodology for high-quality, low-cost design and manufacturing of less sophisticated copiers and duplicators.

Xerox's research organization sought to counter Japanese competencies by offering performance options that could transform low-end copiers to desktop laser printers. The first of these efforts yielded a small, multifunction xerography-based, laser printer/copier device. Delays brought about by commitments to high-end reprographic and printing products and increasing design complexity in the attempt to incorporate more high-performance functionality into the multifunction concept resulted in a product too large and too expensive to compete in the emerging laser printer market.

Continued pursuit of concepts that might recover the corporation's position in this emerging market gave rise to conflict between research and development. Research favored basing low-end printing technology on a new ionographic approach that, because it embodied more electronic manufacturing content, might mitigate the corporation's lack of low-cost design and manufacturing competence. Development favored basing it on traditional xerography. Senior management supported research's approach, and a research-connected "skunk works" development organization was established to develop low-end multifunction printer products. The effort ultimately failed because of the team's lack of product delivery experience, and Xerox's opportunity to participate in initial market development was lost.

The research organization found it difficult to be a significant contributor to the TQM processes that supported the recovery of the corporation's reprographics business. The continuous improvement process in which quality is pursued through requirements stability and conformance to process seemed unnatural to the organization, which viewed its role as one of inventing radically different technologies capable of transforming the industries served by the corporation. Indeed, the research organization sought to generate technologies and capabilities that, if successful, would render obsolete many of the core capabilities that continuous improvement efforts were aimed at reestablishing. Research's perspective and continuous improvement efforts were on target, but unfortunately they were disconnected.

To support investment in the office market outside of reprographics, PARC created an extraordinary set of investment options in emergent workstation and networking technologies. While the research community created a set of technologies as radically new and market forming as xerography had been a decade before, it was not encouraged to take experimental systems to market. Instead, the corporation chose to pursue a leadership position in this new market by investing heavily in the introduction of a set of integrated products and services that it hoped would reinvent the office workplace. This investment was technology focused but failed to consider how the customer would evolve into the technology's use.

The set of products and services offered required customers to make a substantial commitment to a totally new way of working. Creative contributions by customers and third parties were impossible because the set of products and services was closed and proprietary. While the architectural and technical concepts behind the system were valid, the market was developed incrementally by others with implementations that were less technically complex. Thus, Xerox's proprietary, vertically integrated business model missed the open system, horizontal business structuring that ultimately characterized the advance of the distributed computing paradigm in the office workplace.

In 1986, when he became head of corporate research, William Spencer initiated a process of rethinking research's activities and influence on the corporation. He relocated the corporate research office to Xerox's corporate business headquarters in Stamford, Connecticut, became a presence in the senior executive corridors, and inaugurated a process that was intended to bring the research organization into the corporation. Under his leadership, a causal analysis was undertaken of three significant products in which other companies had achieved market leadership on the back of Xerox's early technical leadership (Smith et al. 1988). Among the key contributors to missed opportunities were:

1. Failure to couple marketing and technology.

2. Insulation from rapidly emerging markets because of commitments to existing markets.

3. Failure to be capable in all aspects of the innovation process. This included failure in defining customer interests and needs and failure in recognizing a new or emergent customer base and designing, manufacturing, selling to, and servicing that base.

4. Mismatch between the tempo of external market/business/technology cycles and the tempo of Xerox's internal management and work processes.

The dominating issues were the lack of clarity about research's role and the integration of that role into the corporation's vital commitment processes. Overall, there was a failure to recognize that innovation must be a total business process. It must involve the entire process of technology creation, marketing, design, manufacturing, sales, and support if it is to create products and services for either existing or emergent customer bases.

In the latter part of the 1980s and early 1990s, corporate management recognized that the problems revealed by Spencer's analysis were not restricted to the research organization but systemic (Howard 1992, 107–21). It was realized that innovation during a period of rapid technological change must be a core business process and involve all business functions. For this to occur, research had to play a central role in the corporation's strategic and operational management.

Expanding the Role of Research

For research to play a broader role in any corporation's innovation processes, its leadership and scientists must view themselves in a broader perspective. A research organization within a corporation must measure its successes in terms of its contributions to the growth of the corporation's business value. The research organization must see innovation in terms of the corporation's customers, markets, design and manufacturing operations, product service, core competencies, and learning processes. Because it works in the future time of the corporation, the organization must enable a learning process that refines the inherent uncertainty of the future by working at the cusp of the confluence of markets and technology (see Figure 4-1).

If successful, such a learning cycle effectively immerses research managers in the issues of how to create:

1. New technology that yields a competitive advantage for existing markets.

2. New technology that creates novel competitive options for new markets.

3. Marketable intellectual property that expands capital.

4. Productive innovative processes that expand the capacity for growth.

In 1992, Paul Allaire led the corporation into a new organizational work architecture. Xerox had grown into a highly vertical, functional organization in which power was split between two large functional

Figure 4-1. Emergent Market and Technology Learning Cycles

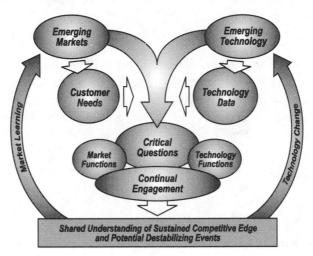

entities: direct sales and marketing; and product engineering and manufacturing. Allaire created an alternative corporate structure that balanced decentralized business divisions with two centralized organizations, research and technology, and customer operations. The design was intended to promote effective links between markets and technologies by according business divisions end-to-end responsibility for products (see Figure 4-2). It creates parallel channels to couple markets and technologies through decentralized highly market-focused divisions. Thus, the design establishes a balance between the advantages of decentralized divisions and the advantage of scale in technology investments, a characteristic advantage of large corporations.

Under the new design, the senior vice president for research and technology became part of the Corporate Management Committee, which was responsible for the corporation's overall strategic and operational management. The senior vice president for research and technology also became responsible for the corporation's technology management process. The research and technology function, which includes research, advanced development, corporate technical architecture, new engineering and product delivery processes, and core competency and new market development, is an integral part of the technology management process. To assure direct and timely perspectives on major technology trends and engineering competency requirements, separate chief scientist and chief engineer positions were established. These positions are jointly responsible to the CEO for consultation and senior vice president for research and technology for their operations.

The Technology Decision Making Board, which is comprised of division presidents, senior research and technology managers, the head of the corporate business strategy office, and the heads of machine and materials manufacturing, is a key governing facility in the technology management process. Chaired by the senior vice president for research and technology, the board oversees the corporation's technology investments and major shared investment decisions and is responsible for establishing a community of senior managers who hold a common understanding of and perspective on technology issues over the entire range of businesses. The board fosters a community of practice among these managers that forms the basis for a common vision and purpose in the corporation's strategic technology investments. In conjunction with the community of senior managers, the heads of corporate research and technology and the corporate business strategy office recommend to the Corporate Management Committee the level of investment for the entire corporation's research, development, and engineering operations over a three-year strategic time frame.

Research and technology investments have become a key part of the corporation's strategic planning process. There are four types of investments in the central research and technology organization. In-

Figure 4-2. Xerox Organization Architecture

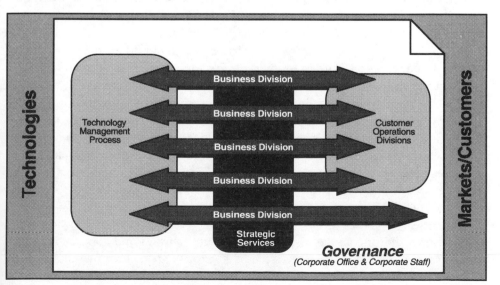

vestment in pioneering research, which is funded by the corporate office and characterized by high levels of uncertainty, has the purpose of discovering emergent technologies that may shape the company's strategic vision and generate future options (Brown 1991, 102–11). Areas of investment are determined by the technical vision of the research management.

Investment in the identification of emergent markets and technologies permits the exploration of opportunities and uncertainties associated with markets that are presently outside the scope of existing business divisions but may expand the company's vision of its role as *the* document company. Protobusinesses are funded by the corporate office and employ technology developed by the research centers.

The third type of investment—investment in strategic capability—has the purpose of establishing new technology platforms and skills that are of recognized importance to the company. Strategic capability is responsive to the Technology Decision Making Board. Major technology investments are shared across divisions and charged collectively to them in proportion to the size of their product development and engineering investments. This type of investment is clear on the possibilities of a specific emergent technology but may be uncertain about its technical feasibility and timing.

The fourth type of investment is in core technology. This is technology that is ready for use in existing or emerging businesses. Investments are contracted with the responsible business entity and charged to the cost of engaging the opportunity. This type of investment, often premised on learning and experience derived from the foregoing types of investment, has the lowest level of uncertainty.

The total innovation process requires research to play a broader role in the corporation's customer, market, design, manufacturing, product service, and learning processes. At Xerox research has gone beyond its traditional role of providing proprietary technology options and technical information specific to the company's business needs to become an active agent in the corporate organizational learning process. Senior research managers are actively involved in the process of forming and executing the corporation's strategic vision. The Allaire senior management team that created the Xerox 2000 strategy included key members of the research and technology organization. The strategy was based on research's reading of the technological forces of change and how these would impact the corporation's future markets and competency requirements. The corporate vision of Xerox as *the* document company has in turn become the research and technology organization's long-range strategic focus. As a result, in order to understand and develop strategic positions that will enable the corporation to offer services in the emerging national information infrastructure, the research and technology organization has led the formation of a Xerox national information infrastructure [NII] policy committee that is composed of Paul Allaire, members of research management, and other members of the Corporate Management Committee, corporate strategy office, and government affairs.

Under the total innovation process, a codevelopment team works directly with a lead customer to reach a shared understanding of an important problem that may be solved by an emergent technology and then develops with the customer a solution for the problem (Brown 1991, 102–11). Lead customers are engaged on the basis of the learning opportunities they represent to the company's market/technology vision. The nature of the problem and the quality of an enterprise determine the magnitude of the learning opportunity for advancing the state of the art of document processes and the relationship of these processes to productive work practices. The first codevelopment team was composed of people from research, development, and marketing and led by PARC. The team worked with the manager of case study printing and distribution in the Harvard Business School Publishing Division to devise an electronic document production management system for printing business cases on demand and to integrate existing, valuable paper-based case study collections into the system.

A similar team of people from development and marketing with support from research was formed to address a document problem of national, indeed, international, importance. Books published since the mid–1800s are decaying at an alarming rate because of the acid content of their paper. The Xerox team worked with the Cornell library to develop a system that would enable books to be scanned, browsed electronically, and printed remotely on demand. A version of the system is being used by the Cornell bookstore to prepare course packets (collections of readings and notes) for students.

Both of these projects moved the corporation beyond its previous state of competence in document management solutions. New technologies related to document recognition and conversion were drawn from PARC, image processing technologies were drawn from WRC. New software prototyping tools were developed, and the notion of an extensible document solutions architecture was presented.

As a part of the total innovation process, a strategy of radical incrementalism has been adopted. Visionary concepts are developed and tested in the marketplace through small, market-learning events and the acquired learning is incorporated into the next application (see chapter 9). For example, ubiquitous computing, a powerful vision developed at PARC in 1990, is premised on the anticipation that computational resources will be embedded in a variety of interactive surfaces that will be available wherever one goes through a system of wired and wireless connectivity. A dynamic system that supports ubiquitous access to electronic information, document services, and computational resources independent of location, includes document activation via video, voice, and animation. Physical artifacts of the environment include: PARCTabs, lightweight, pocket-sized computational display devices with reminder, calendar, and locator level services; PARCPad, a portable, notebook-sized, paperlike universal interface that provides a personal work surface connected to the computational and communication infrastructure; and Liveboard, a white board-sized interactive work surface connected to information and computational resources and offering remote collaboration services. Elements of this system have been built and used at PARC in much the same way that the Alto personal computing environment was built and tested a decade earlier. This time, however, a different approach is being taken to the marketplace. Instead of building and introducing the entire system, research is exploring selected elements in the marketplace. The Technology and Market Development Group, a part of the research and technology organization, has been chartered to create protobusinesses to explore customer uses of new technologies and the business structures needed to support market growth. Liveworks, a new business that has taken the Liveboard part of the ubiquitous computing concept to the marketplace, is run as an internally funded venture in the venture capital model. If successful, it may become a new division or spin-off. Through customer participation and market feedback, research is learning how this technology will enable new collaborative work practices.

The research organization also helps to condition the corporation for change and the restructuring of work practices within the corporation. It can be particularly helpful in delineating how technology may enable productive new work processes. Three applications of emergent technology are intended to improve productivity and time-to-market for Xerox's internal operations. The applications are the use of remote personal communication devices to support field service personnel, the use of increasingly available, cost-effective mips and bandwidth to support productive design processes, and work process redesign to reduce radically time-to-market for materials design and manufacture.

Xerox's Denver district sales and service organization has long had a reputation within the corporation for innovative and productive work process reengineering. John Seely Brown, chief scientist and PARC center manager, visited the head of service in Denver to offer to fund the use of portable communication devices to support field-based technical representatives, provided PARC scientists were permitted to study the impact of the technology on formal and informal work processes. The technical representatives are key customer contacts and their jobs represent an important core competence of the corporation. Their responsiveness to customer calls and speed of repair are major determinants of customer satisfaction. They work at a distance from one another, directly with customers, on many different products based on continuously changing technology and widely varying service requirements. The PARC study found that although their formal processes were well documented, it was the informal processes that were most active in their practice.

The use of personal portable devices had a significant impact on repair time. The devices enabled technical representatives to share experience in real time and support one another in their work. The informal, "coffee pot" learning processes so critical to technical representatives were supported by the remote communication technologies, which will become part of the general service practice.

What will be the impact of unrestricted amounts of computation and communications power on the productivity of engineering design processes? This is the question the second application of emerging technology to Xerox's internal operations seeks to answer. What impact will having 2000 or more mips at their disposal have on Xerox engineers' work practices? How will unconstrained bandwidth affect collaborative work distributed across time and space? What may be the human machine interface barriers to rapid diffusion of this power into the work processes of organizations? These questions are being addressed collaboratively by Xerox's Wilson Research and Technology Center [the successor to WRC] and the Design Research Institute at Cornell University's Center for Theory and Simulation in Science and Engineering. The intent is to apply the computational power present in today's supercomputers to Xerox engineering issues. It is presumed that such power will soon be available to individual work environments. Real-time computational prototyping enables designs to be simulated, thereby reducing the need for the time-consuming development of hardware or chemical prototypes. The simulations become part of the corporate design memory. Thus, the design experience becomes recallable and reusable and can act as the basis for the development of tools and methods to support the sharing of engineering information and the creation of shared engineering work spaces that facilitate remote collaboration.

Process reengineering aimed at making research itself more productive and time-to-market responsive is the focus of the third application of emergent technology. For the corporation, the design and worldwide manufacture of xerographic materials and chemicals are important core competencies. The materials research center in Canada (XRCC) has speeded materials development by organizing teams of chemists and chemical engineers to work collaboratively on synthesis. Reactions are chosen at the early stage of research with a concern for scale of production. The center has developed a state of the art pilot plant facility that is premised on the notion of "scale down" rather than "scale up." Experimental reactor size is based on the ability to move rapidly to economic manufacturing. The benefits of the approach are manyfold. The coordinated development of materials innovation in concert with host hardware development is greatly enhanced and the greater confidence in manufacturability early in development permits exploration of many more materials innovations in new products.

CONCLUSIONS

Recognizing that the ability to link technology and markets is essential to its business prosperity, Xerox is making innovation a core business process. To do so, the corporation has made the research organization an influential, active participant in the company's strategic and operational management. The risks associated with this change in the role of the research organization must be weighed and carefully considered. Bringing research into the central processes of the corporation risks rendering it less independent and thus muting the voices that have pronounced the need for radical change and have proclaimed so visionary a future. It risks altering the shared management values that have made the research organization so prominent in scientific communities. It risks underutilizing well-developed areas of competence by asking scientists and engineers to engage in fields of knowledge outside their expertise. At the same time, it offers the research organization a new opportunity for thinking about problems in unconventional ways. It affords research an opportunity to contribute to Xerox's determination to remain one of the most innovative and productive corporations in its chosen fields of business by learning to thrive in an environment of change and uncertainty.

References

Brown, John Seely. 1991. "Research That Reinvents the Corporation." *Harvard Business Review* (January–February 1971): 102–11.

Desseaur, John H. 1971. *My Years with Xerox: The Billions Nobody Wanted*. Garden City, New York: Doubleday.

Garfield, E. 1993. "Citation Index for Scientific Information." *Science Watch* 4, no. 2: 8.

Howard, Robert. 1992. "The CEO as Organizational Architect: An Interview with Xerox's Paul Allaire." *Harvard Business Review* (September–October): 107–21.

Jacobson, G., and J. Hillkirk. 1986. *Xerox: American Samurai*. New York: Macmillan.

Kearns, David T., and David A. Nadler. 1992. *Prophets in the Dark: How Xerox Reinvented Itself and Beat Back the Japanese*. New York: Harper Collins.

Pake, George. 1985. "Research at Xerox PARC: A Founder's Assessment." *IEEE Spectrum* (October): 54–61.

Perry, Teklu, and Paul Wallich. 1985. "Inside the PARC: The Information Architect." *IEEE Spectrum* (October): 67–75.

Smith, Douglas K., and Robert C. Alexander. 1988. *Fumbling the Future—How Xerox Invented, Then Ignored, The Personal Computer*. New York: William Morrow.

Smith, T., W. J. Spencer, T. J. Kessler, and J. Elkind. 1988. "Why New Business Initiatives Fail." Xerox Corporation Study, Stamford, Connecticut.

Uttal, Bro. 1983. "The Lab That Ran Away From Xerox." *Fortune* 108 (5 September): 97–102.

5

REINVENTING RESEARCH AT IBM

John Armstrong

A LTHOUGH ALL OF US are deeply convinced of the importance of industrial research, I expect that we also all agree that success in R&D is not enough to guarantee the success of individual firms. This remark is an important part of the context for any discussion of the future of industrial research. It reminds us that we must always keep in mind the many factors other than R&D that are necessary for success and that, if done in a first-class way along with effective R&D, will be sufficient to assure success. We must also remember that it is very dangerous to generalize across industries. What is true of R&D in the chemical industry may have little to do with current issues in the R&D of the electronics industry. Moreover, large, multinational, multi-product firms are often engaged in different industries, a fact that complicates their management of R&D. For example, the IBM Research Division serves about a dozen different internal customers who work in industries with very different development cycles, barriers to entry, and maturity. Finally, we should recognize that there is no textbook model of industrial research, no "one size fits all" prescription. The IBM Research Division has been consciously engaged in reinventing itself for at least the last twenty-five years. As director of research, I believed I had two main goals. One was to make sure that the organization did first-rate scientific and technical work. The other was to make sure that good work mattered to the corporation. This goal requires constant reinvention of the research organization itself.

I propose to describe some of the things that we have done and are trying to do to make the IBM research division increasingly effective despite changes in our industry and in our business. Many of these efforts parallel those discussed by Mark Myers in chapter 4. We too find it useful on many occasions to blur the distinctions between research and development and manufacturing. We too find it important to remove internal competition between research groups and development and manufacturing groups. Moreover, we have found all of these efforts to increase the effectiveness of our research organization are complicated by the fact that the competitive situation is changing more rapidly now than ever before, which means there is less time in which to adapt to change.

First, I want to describe what are called joint programs in the IBM R&D enterprise. Over the last ten years or so, the IBM research division and its sister divisions have created a series of programs in advanced technology and early product development. These programs involve work that is jointly planned, staffed, and funded by the research division and the appropriate product divisions and laboratories. There are joint programs in semiconductor technology, in the design of scalable supercomputers, in database technology, and in dozens of other areas. These have succeeded to the extent that, in some cases, there are members of the research division who work in development labs for extended periods, and conversely, there are groups of developers on assignment to research or in groups managed by colleagues from research.

The joint programs came about as a response to the competitive conditions of the 1980s. In the 1990s, we have all had to learn new lessons, and one of the most important lessons we have had to learn is that the utility of a particular research portfolio changes over time. After thirty years of progress at a rate of 25 percent per year and after internationalization of the know-how required, many of the hardware technologies in computing and telecommunications are available as commodities. Therefore, these technologies in themselves are no longer the source of comparative advantage that they once were. For decades, we have invested in two major semiconductor technologies, bipolar and CMOS. Years of exponential progress in CMOS now make it possible to do with that technology what was previously possible only with bipolar technology. Furthermore, in pushing forward in CMOS, we have formed a three-way alliance with Siemens and Toshiba, with the work being done in Fishkill. This halving of the technologies and the three-way sharing of the investment imply, for IBM at least, that the investment in semiconductor R&D is several times less than it was eight or ten years ago. Painful as such portfolio realignments are, they are required of many firms these days.

In addition to internal joint programs, IBM's research division also participates in intercompany alliances. There is, for example, a joint venture between IBM and Toshiba that is called Display Technology

Inc. It was formed to develop and manufacture active-matrix, flat panel liquid crystal displays. If you've seen the new IBM 750C laptop, you have seen the most recent fruit of that successful joint venture. Researchers from the Watson Research Center made substantial contributions to the display technology and to the pointing-stick that is integrated into the keyboard and replaces a mouse.

Another exciting development has been the proliferation of joint projects between researchers and customers. A group of our computer scientists worked with the marketing division and the State of California to develop an integrated records management system for the Division of Bankruptcy of the State of California. When a person or a firm files for bankruptcy in California, all of the state's departments want to know instantly in order to make any appropriate claims. This turned out to be a fruitful collaboration, as well as a good education for the researchers, that resulted in a successful IBM product.

In another initiative four or five years ago, research management went to corporate management and asked for authorization for fifty additional employees but did not request money to pay for them. Some of you will understand that this was an unusual request. We proposed to develop product software in what was then, for IBM, a new way, and we proposed to fund the work by going to development labs and offering to do for one-third or one-fifth of *their* projected costs, projects they felt they could not afford to do. We set up small teams with excellent tools and support and gave each team end-to-end responsibility for the product code—from design to test. It has worked very well, and there are now hundreds of thousands of lines of code in IBM products that have come out of what has become the Experimental Software Development Center. Even more important, our experience in the ESDC has helped to propagate the paradigm of product development by start-to-finish, highly motivated small groups.

I would like to mention two other things. For a long time, the research division did not understand how to work directly and effectively with manufacturing. However, we have recently found that we can exert tremendous leverage if we go at it in the right way and choose the right set of topics. Some of our manufacturing divisions now acknowledge that hundreds of millions of dollars of savings per year result from work that has come directly from the manufacturing research group of the research division.

Another initiative that is exciting to the research division and shows promise of being a new way to be effective is the formation of internal spin-off "companies" from the division. The first internal spin-off created a 32-way parallel super-visualization computer. It was aimed at the scientific data visualization market, but to the surprise of researchers, the principal market has been the computer animation and special effects sector of Hollywood. It is by far the biggest customer. The very high performance visualization system was developed and brought to market in two years with a team of only fifty people and a "board of directors" made up of senior IBM management, primarily from outside the research division.

Finally, we have decided that it is a very good idea to make external spin-offs as well. Probably none of us at this conference is old enough to have to worry about hip replacements, but I predict that if one of us does have a hip replaced any time after the next few years it will be done with the help of "Robo-Doc," a surgeon's robotic assistant. "Robo-Doc" is a product of a spin-off from the manufacturing research operation in collaboration with doctors from a California hospital.

In summary, then, one of the keys to the future of any industrial research organization is a willingness to reinvent itself. There is no textbook to follow. In this endeavor of institutional reinvention, the research culture that young scientists bring to industry from their universities is almost entirely useless. We spend a lot of time teaching our people how to think straight about science and technology and their relationship to industry. It is only when this is understood that our researchers can help to renew the corporation and reinvent what corporate research itself means.

6

THE FUTURE OF INDUSTRIAL R&D, OR, POSTCARDS FROM THE EDGE OF THE ABYSS

Peter R. Bridenbaugh

MOST SPEAKERS ARE FACED with the challenge of dreaming up some catchy title that will provide insight into their remarks and bring a few folks into the hall. Even when they have been successfully creative, few speakers I know start with the title or feel particularly obligated to stick to the subject it represents once the room is filled. However, I was so struck with what I perceive to be the underlying tone of this symposium—that industrial R&D is in serious danger of going over the brink—that I am actually going to begin with my title, "The Future of Industrial R&D, or, Postcards from the Edge of the Abyss." The title alone puts my views in perspective. While I am not certain whether we are about to slip into a chasm of chaotic decision making around science and technology or whether we are simply responding to changes in the external competitive realities that have shaped the size and scope of industrial R&D since its inception, I can look at the trends I see in Alcoa and American industry today and see the potential for a precipitous fall into an abyss that is shrouded by short-term, reactive thinking. The data show that industry continues to fund the majority of R&D in this country. Indeed, if I read the latest NSF numbers correctly, industry funded some $81 billion of R&D last year. Even though 1992 marked the first increase in industrial spending after three years of declines, I remain somewhat skeptical of this number. It may reflect clever corporate accountants becoming skilled at taking advantage of R&D tax credits more than an increase in meaningful industrial R&D.

One's feelings about an abyss clearly depend upon where one is standing. If the abyss lies in front of you, your sense of well-being, of continuity, and of the future is quite different than it is if the abyss lies behind you as an obstacle already overcome. To my mind, the question we need to ask is not whether the abyss lies in front of us or behind us but whether we are even facing the right direction. I'd like to begin to ferret out the answer to this all-important question by looking back at Alcoa's history. Perhaps by seeking the lessons of the last 105 years and seeing if we can apply them to the current era—to external competitive realities and to internal corporate thought processes and decision-making philosophies—we can determine not only the future of R&D in Alcoa but the future of the enterprise itself.

My view of the future continues to evolve. It is quite different today from what it was six years ago. My perspective has responded to subtle changes and dramatic upheavals in the internal character of my company and in the global environment that appears to be redefining the nature of economic competition on an almost daily basis. I believe that industrial R&D institutions have always existed to serve the competitive needs of the parent company in any given era. How well this mission has been achieved has been as dependent upon the capacity of corporate leaders to understand these needs as upon the capabilities of the R&D institution itself.

What I would like to do is to share with you my view of the challenges facing industrial R&D labs in the 1990s and early part of the next century and how one company, Alcoa, is approaching these challenges. I would like to step back to 1919 and begin the journey forward through the eras of Alcoa's scientific and technical history to see what may be gleaned from the company's responses to the competitive realities it has faced. Our reactions were not always right, but over time, I think we have a pretty good track record in the industry that we founded.

Before I delve into our history, I would like to give you my view of the role of industrial research. To my mind, the primary role of industrial research labs is to acquire knowledge, either internally or externally, and translate it into processes, products, and solutions that create competitive advantage for the enterprise and commercial success for the shareholders. It is important to emphasize this point because I sense that much of the concern expressed here about the looming demise of industrial R&D revolves around its role in basic, or fundamental, research. I would argue that this type of R&D has never been the function of industry, with the exception of Bell Labs and a handful of others. In the early days of the Aluminum Research Laboratories, its so-called pioneering work was truly applied research, highly focused on the particular problems of a still embryonic Alcoa. Industrial researchers are primarily integrators of science and technology. They are professionals whose function is to bridge the gap between basic and applied research, between science and technology and real-life products and processes. This role mandates a brand of R&D that I refer to as industry-directed fundamental work. This is vital work that must be protected.

The creation of the Aluminum Research Laboratories (ARL) in 1919 initiated formal R&D at Alcoa under the leadership of Dr. Francis Frary. Its creation was driven by the realization that Alcoa lacked sufficient technical understanding to meet the emerging materials needs (specifically those of aerospace) of the nation. This had become apparent during World War I. Dr. Frary was asked to make Alcoa the world's premier technical company by assembling a cadre of world-class scientists to build a strong foundation in fundamental understanding focused on the technical problems in electrochemistry, chemical engineering, physical metallurgy, and analytical chemistry that confronted Alcoa at that time. His work was assisted by the government, specifically the U.S. Navy, which worked closely with Alcoa to develop new alloys for aircraft. By the time this era ended in 1949, it was apparent that Dr. Frary and his colleagues had succeeded marvelously.

The next era began in 1950 and ended around 1965. It was sparked by the presence of domestic competition in the aluminum industry for the first time since Alcoa's inception and by the pent-up demand for consumer durables created during World War II. In Alcoa's view, the time had come to "aluminize the world." Application engineering became the major task for the R&D organization. The storehouse of knowledge built by Frary's organization became a resource to be guarded and exploited but not replenished. Engineers who had bachelor of science degrees were perfectly suited for the task at hand. They were recruited to replace the retiring generation of world-renowned Alcoa scientists. Again, the R&D organization succeeded marvelously, but in so doing, it failed to replenish or advance its fundamental knowledge base.

The next era, from the mid-1960s to the early 1980s, was created by the success of our efforts to aluminize the world, or at least the beer and beverage container industry. New aluminum applications, specifically the aluminum can, were perceived as having the capacity to consume the available aluminum supply for the foreseeable future. About half-way through this era, however, the first rumblings of energy and environmental issues began to be heard at Alcoa. The company's clarion call became process improvement, and the major project of this era was a highly secretive, low-energy smelting process. Application engineering was dismantled as new product development was abandoned. If they thought about it at all, corporate leadership and isolated R&D management considered Frary's storehouse of scientific understanding to be sufficient to see Alcoa through. By the mid-1970s, it was apparent that Alcoa's R&D organization was failing to meet its objectives. The signals were vague at first but became clearer by the late 1970s, when a series of new corporate faces began showing up to "fix" the labs. They succeeded in confirming the obvious but not in understanding the causes of the problems.

The next era began in the early 1980s and, I believe, ended somewhere around 1988 or 1989. It was now apparent that Alcoa had exhausted its foundation of scientific knowledge and had often engineered solutions to problems that were not fully understood. The highly touted Alcoa Smelting Process, for example, had failed. This failure was compounded by significant overcapacity and, as a result of decisions made about R&D directions fifteen years earlier, no meaningful new products. There were now frequent and deep troughs in Alcoa's business cycle. The leadership of the laboratory took it upon itself to define a new R&D charter that was not a great deal different from the charter given to Dr. Frary more than six decades before to build the knowledge base, apply it to new products even beyond aluminum, optimize processes, recruit the best minds available, and return Alcoa to the center of the world's technical stage. While corporate leaders never formally stated their agreement, they embraced these new directions not only with words but with dollars.

This brings us to the present era. Is this era a time of reassessment or a response to the radical changes in direction that occurred in the mid-1980s? My sense is that the answer to this question is both yes and no. I believe there is a significant portion of corporate management that thinks the initiatives of the 1980s were too ambitious and perhaps unnecessary. For these people, today is indeed a time of retrenchment and reduction in technical scope and effort. In the minds of others, however, it is a time to channel our knowledge and efforts into product and process development, just as the company did in the 1950s and 1960s. These people seek to harvest the knowledge that has developed in the last ten years. Neither school of thought is new nor unique in Alcoa's history.

What have we learned from our past (other than the accuracy of Santayana's conclusion that history repeats itself) that can be applied to keep us from a chaotic and abysmal future?

- First, and perhaps foremost, we have learned that R&D in Alcoa is a reflection of the interplay between the vision and judgment of its leaders and their response to the external pressures confronting the company. To leverage financial and human resources committed to technology, we must realize the vital importance of vision, judgment, and leadership from the top of the corporation. The power and influence of the chairman is especially critical. The ability of the CEO to promote technology as the engine of change and to maintain critical technologies at the leading edge in order to shape the future even as external competitive realities continue to press for short-term performance determines the true effectiveness of technology.

- We have also learned that our R&D organization is robust and reactive to the vision and judgment of its corporate leadership but only when it is vitalized by the capacity to access and understand the base of fundamental scientific knowledge in areas critical to corporate direction.

- As a member of corporate leadership, I hope we have learned that there is a price to be paid for abandoning the pursuit of fundamental understanding. We have seen in the 1930s and again in the 1980s the necessity of remaining in the midst of scientific advancement, especially in times of revolutionary progress. If we will just allow ourselves the luxury of dispassionate observation, we will see the all-too-imminent chasm facing any corporation that is technically isolated and incapable of harnessing scientific progress because it cannot recognize and apply it.

- We have learned that there are few good reasons to abandon product development in the face of short-term pressures, that it takes a decade or two to feel the full impact of such a decision, and that it is then difficult to rebuild this capability.

- We have learned that the government can be an effective partner under certain circumstances. Our development of aerospace alloys would not have progressed as it has without the support of the government. We cannot discount the efficacy of well-managed public-private partnerships or the waste of poorly conceived technology initiatives.

It seems apparent that the ascent of Alcoa and its scientific and technical community to new heights or its descent into an abyss of mediocrity and eventual extinction is going to depend upon how well we can match the insights of the past against the competitive realities of the present and future. Personally, I feel we have somewhat of an advantage over some other companies because we have at least tried to understand the causes and effects of the past and learn from them.

I would like to recast these lessons learned in terms of the traits that are requisite for effective corporate leadership today:

- Long-term vision derived from an awareness and fact-based understanding of external competitive realities on a worldwide scale.

- The ability to lead change, to articulate and communicate clearly its rationale and results to corporate stakeholders from the plant floor to the boardroom.

- The willingness to take risks.

- The ability to create commercial advantage by leveraging technology resources.

If these are critical internal skills for the coming era, what are the external competitive realities against which they will be matched? It is clear that the coming era will be characterized by:

- Three dominant world markets, the EC, North America, and Asia, with higher growth rates for aluminum in the EC and Asia than in the North American marketplace which is dominated by the United States.

- Global dispersion of manufacturing resources.

- Creation and acquisition of technology with no regard for national borders.

- Economic and technical parity between global industrial competitors.

- Diverse social and legal philosophies concerning the role that governments should play in protecting indigenous industries. A worldwide free market system, if one ever existed at all, is simply implausible.

- A growing requirement for environmentally acceptable products and processes.

- Geometric growth in human intellectual progress, made most apparent by the transparent and instantaneous flow of information and our ability to access and share pertinent scientific and engineering advancements regardless of origin.

- Continuing political instability for isolated regions of the world.

It is these realities that will underpin corporate decision making and thus shape the size and nature of industrial R&D in the years ahead. Drawing from these external competitive characteristics and from the internal skills that I see as essential for Alcoa or any industry, here are the most likely scenarios for industry and industrial R&D for the remainder of the 1990s and the first years of the next century:

- Global partnerships will be brought about primarily by the desire for a presence in the three dominant world markets within which the industry hopes to play. These partnerships will be based largely upon the technical acumen of the players. For materials suppliers, those possessing the scientific know-how to integrate materials-process design and product design will be sought. Companies perceived to be technically isolated or obsolete will not be players. Mutual profitability, of course, will be the foundation for these partnerships. The sharing of risks and rewards will be essential.

- Transparent information flow and accessibility based upon worldwide open system networks will be commonplace. Industrial R&D will become multicultural, multilingual, and multilocational.

- A much clearer definition of the roles of all R&D performers in this country will come about as a result of the complexity of the problems to be solved and the costs of doing R&D. I believe we will see new roles, primarily in areas of standards, characterization, research tools and supercomputing, for federal laboratories while academia will retain its emphasis on fundamental research.

- Companies that are successful will continue to keep a balance between industry-directed fundamental and applied R&D. This is the bridge between fundamental and applied R&D that I alluded to earlier. It represents what I view as a crucial link between academic, governmental, and industrial R&D. Industry-directed fundamental work is conducted in parallel with basic research to assure the rapid transfer and implementation of fundamental knowledge into product or process development programs. This work requires a cadre of scientists with doctorates who can hold their own with their counterparts in basic research. I firmly believe that people cannot hear what they do not understand. We must be able to understand the opportunities that progress in basic research may present to us. If we cannot, we will not remain competitive.

- Industry will gravitate toward the concept of core competencies, or strategic technologies, to focus its efforts. The process for identifying those technologies will impact R&D by targeting diversification strategies, and by defining the role that centralized laboratories play vis-à-vis business-specific technology organizations. Properly used, core competencies will help to determine how and where industry should fund basic research and where it can look for commercial and technical linkages.

- Research will be conducted through global networks of people, machines, and computers. I can easily envision working on-line with Boeing designers as we simulate the performance of different materials in a new wing design or running crash simulations with Audi based on various combinations of product geometries and materials characteristics. As I see it, the only reason we will not move to these kinds of collaborative research programs will be technical illiteracy and isolation. If there is indeed an abyss for industrial R&D, this is it.

What, then, must those of us in industrial R&D do to negotiate a safe crossing of the abyss that lies before us? Based upon what I have learned from Alcoa's past and from what I see is required of current corporate leadership, here is a short but on-going "to do" list.

- We need to educate senior management, even CEOs, about the importance of a strong internal technology base and how to leverage it. Here is an excellent opportunity for those of you in business schools, particularly schools associated with engineering schools. We need a far better way than we presently have to create among today's business leaders awareness of the leverage that technology can provide. This may be viewed as a continuing education issue as I do not believe we are going to see recent MBAs in executive suites for another ten to fifteen years. It seems to me that we must also emphasize the reality that global partnerships are essential and that technical leadership is a basic requirement for being perceived as a valued partner.

- We also need to educate senior managers about the symbiotic relationship between basic research and industry-directed fundamental work. Far too many senior managers fail to see the distinction between these two categories and think of them both as a kind of barking at the moon, "fine for academics but not on my time." I believe we have got to make it clear that industry-directed fundamental R&D and the skills required for it are essential if we are going to be able to harvest the world's ever-growing knowledge base.

- With apologies to any of you who believe money is not a motivator, we need to design compensation systems that reward both short- and long-term performance far better than do the systems we have today.

- We have to find new and creative ways to use university and national laboratories effectively. Here is where we must use all the information and computing tools at our disposal to minimize costs and maximize cooperative effort. Conducting research in software, linking scientists and engineers through high-speed networks, and making information readily accessible are key enablers, particularly if the needed rationalization of research universities and national laboratories occurs.

- Finally, industrial scientists must be willing to redefine themselves without compromising what they know is right. More and more, our work is becoming focused on bridging the gaps between fundamental knowledge, applied research, and development. Industry-directed fundamental work is essential to this role, and we cannot abandon its pursuit under the pressure of short-term problem solving.

If I am correct in my assessment of today's trends, the need for aggressive, credible, and persuasive leadership from senior industrial R&D management is critical. Some of us may have to throw ourselves on the track to stop the technical service train from carrying our organizations to ignominy. We cannot simply roll over as we did in the 1950s; the pace and value of scientific and engineering progress simply will not allow it. In today's era of global competition, of dispersed manufacturing and technical resources, of parity in intellectual capabilities, those in industrial R&D need the instincts, the acuity, and appetites for knowledge that hungry wolves have for their prey. These are the survival skills for industrial R&D in this era of global competition. They are the skills that will allow us to break a trail across the abyss that others may follow.

7

SOME PERSONAL PERSPECTIVES ON RESEARCH IN THE SEMICONDUCTOR INDUSTRY

Gordon E. Moore

IN AN INDUSTRY in which survival hinges on conducting research, Intel Corporation is distinct among semiconductor manufacturers in that it has no formal research organization and yet it invests steadily in R&D. A review of the industry will help to elucidate this seeming contradiction.

AN INDUSTRY PREDICATED ON R&D

The semiconductor industry was a $1 billion business in 1960. Today, its revenues exceed $100 billion and it is the basis of a $1 trillion electronics business worldwide. Product cycles are very short in the industry, and each new generation of semiconductor products and technology completely wipes out the previous generation. To appreciate the level of cost effectiveness that accounts for this phenomenon, an analogy to real-estate development might be helpful. In real estate, the higher the density of development, the more valuable the property. From the beginning, the selling price for silicon has been $1 billion per acre. At first the industry could place a transistor on one-quarter of a nanoacre. Today, it can pack 1,000 transistors into that area. The net result is that prices continually go down as yields go up and the industry gains experience. Because a recession in the industry or the economy as a whole lasts the length of a product generation, companies never recover in existing products. If they come out of a recession at all, it is on the back of new products.

Clearly, then, a successful research and development program is essential for survival in the industry, and the industry routinely invests 10 to 15 percent of revenues in R&D. A plot of Intel's financial performance over the years would show revenues dipping here and there, earnings fluctuating wildly, and R&D expenses following a smooth, exponential growth curve.

The genesis of the industry was Bell Laboratories' invention of the germanium transistor in 1947. Although subsequent work by the lab pointed to silicon as another potentially suitable semiconductor material, the first silicon transistor was introduced (a year earlier than expected) not by Bell Labs or by a vacuum-tube company involved in amplifiers and switches, but by a geophysical company that provided oil well services—Texas Instruments. What may be the next major milestone in the evolution of the semiconductor industry occurred at an aerial survey company, Fairchild Camera and Instrument. Fairchild invented the planar technology that provided the basis for the integrated circuit. The integrated circuit is a complex product with a complex lineage. At the very least, Texas Instruments and Fairchild shared in its siring.

Fairchild Corporation can be viewed, at least in part, as one result of William Schockley's decision to leave Bell Laboratories in 1955. Seeking success in business as well as in science, Shockley, the inventor of the transistor, established his own laboratory in Palo Alto, California. He proposed to batch fabricate a silicon transistor based on technology Bell Labs had placed in the public domain because of a consent decree by the Justice Department. To do so, he recruited a number of chemists and physicists. As the physics of the devices was not well understood, the technology was not at a stage to warrant engineers. Hence, the work involved fundamental research aimed at understanding the technology and device structures. Eventually, a set of problems that hampered laboratory activities led a group of lab workers to seek support elsewhere. The group decided to collaborate with Fairchild Camera and Instrument, which subsequently founded Fairchild Semiconductor Corporation to continue work on a batch produced silicon transistor.

Fairchild Semiconductor's timing and direction were extremely fortuitous. Semiconductor science and technology were evolving rapidly. Indeed, the technology led the science in a sort of inverse linear model. (In fact, it was once seriously proposed that Halloween should be designated an industry holiday as so much witchcraft was involved in processing.)

In the planar structure, Fairchild struck a rich vein of technology. Worth noting in connection with Fairchild's development of the planar structure is the impact exerted by defense R&D. Contrary to what may be thought or alleged, the space program in the 1960s had a negligible impact on the semiconductor industry. About the only defense spending that did have an appreciable impact was spending for the Minuteman I and II missiles. Refinement of the planar transistor was probably hurried by the Minuteman I, and Minuteman II supplied the first volume market for integrated circuits. After 1962, this influence evaporated, and the military has not had an impact on silicon product development since. In any case, integrated circuits, MOS transistors, and the like proved too rich a vein for a company the size of Fairchild to mine, resulting in what came to be termed the "Silicon Valley effect." At least one new company coalesced around and tried to exploit each new invention or discovery that came out of the lab. Notwithstanding spinoffs in all directions, there was still plenty to keep Fairchild Semiconductor growing as rapidly as it could.

New technology was plentiful, but its scientific underpinnings were largely obscure. Chemically based electronics functions had slipped between the cracks in university research. Because it fit neither the chemical engineering nor electrical engineering departments of universities, for a decade basic research in semiconductors was the province of the semiconductor industry.

Fairchild Semiconductor invested in excess of 10 percent of its revenues in what grew into a 600-person, stand-alone R&D organization. As it was in the right technology and functioned quite effectively, the laboratory was highly productive for a time. Then, in the late 1960s, it began to have difficulty transferring new products and technology to the product and production divisions. In 1968, for example, Fairchild still had not transferred to production MOS transistor devices that it had had stable in the laboratory since 1961 even though the technology was being exploited successfully by companies Fairchild had spun off and by spin-offs of those spin-offs.

It would seem that the more technically competent a receiving organization becomes, the more difficult it is to transfer technology to it. When it clearly had all the technical capability in the company, the Fairchild lab had no trouble transferring technology to production. As the production organization became more successful and began to recruit more technical people, however, technology transfer became more difficult. Production, it seemed, had to kill a technology and reinvent it in order to get it to manufacturing.

INTEL: AN EXPERIMENT IN COLOCATION

Turmoil among corporate management at Fairchild led Robert Noyce and this author to found their own company, Intel. In light of recent experience at Fairchild, it was decided to forestall problems with technology transfer by establishing Intel without a separate R&D laboratory. At some cost to manufacturing and probably to R&D efficiency, development would be conducted in the manufacturing facility. Intel recruited a number of the best Ph.D.'s available and spread them throughout the organization. It also established and maintained contacts with major universities that were engaged in areas of research of interest to it. The company continues to follow this course, deeming the time-to-market issues associated with technology transfer to be of paramount importance.

Intel operates on the Noyce principle of minimum information: One guesses what the answer to a problem is and goes as far as one can in an heuristic way. If this does not solve the problem, one goes back and learns enough to try something else. Thus, rather than mount research efforts aimed at truly understanding problems and producing publishable technological solutions, Intel tries to get by with as little information as possible. To date, this approach has proved an effective means of moving technology along fairly rapidly.

With a product as complex as semiconductors, it is a tremendous advantage to have a production line that can be used as a base for perturbations, introducing bypasses, adding steps, and so forth. Locating development and manufacturing together allows Intel to explore variations on its existing technologies very efficiently. It does not, however, accommodate dramatic change. Whenever Intel has explored a totally new technology, such as bubble memory, it has set it up as a separate organization.

There is another advantage to operating on the principle of minimum information: the company generates few spin-offs. Because it does not generate a lot more ideas than it can use, Intel's R&D capture ratio is much higher than Fairchild's ever was. In fact, because the semiconductor business has become sufficiently capital intensive to discourage start-ups, spin-offs generally are less of a problem than they used to be.

Although its approach to R&D may be unconventional, Intel spends quite large sums on it. Current R&D spending exceeds $1 billion, of which roughly one-third goes to process improvement and two-thirds to product development. The company tries to apportion these dollars in round numbers to general areas. Each product group is required to submit a project list ordered in decreasing priority, explain in sometimes excruciating detail why the list is ordered as it is, and indicate where the line ought to be drawn between projects to work on and projects to put off. This notion of striving for efficiency within a prescribed time frame is a marked distinction between Intel and the early Fairchild Semiconductor. When the senior laboratory staff got together after Fairchild put the first integrated circuits into production, the tenor of the conversation was "Okay, we've got integrated circuits; what will we do next?" While operating as it does may, at some point, cause Intel to miss a revolutionary idea that has the potential to wipe out established positions, having a large, competent R&D organization has not been shown to be protection against change in a basic business paradigm. Morcover, Intel's R&D model is not pure. The company does have a small group that tries to stay abreast of what is going on more broadly in the semiconductor industry. This is not, however, the sort of group that undertakes programs from which results are expected in ten years, although Intel has done some of that. Involvement with some of the first signal processing chips in the 1970s led the company to move too quickly to a product. As a result, Intel was selling complete speech recognition systems from production lines at the rate of two per year. It was the wrong business for Intel, which wisely abandoned it.

The company has found a niche in being a supplier of building blocks to the semiconductor industry. It no longer maintains a significant speech recognition activity internally because a leadership position in this technology is not sought at this time. When it becomes appropriate to have special purpose silicon for speech recognition, however, Intel will probably be a producer. Intel's president and CEO, Andrew Grove, promulgates a vision of what this technology will look like about five years down the road, but beyond that, the company's corporate vision becomes fuzzy.

SOURCING BASIC R&D FOR SEMICONDUCTORS

The semiconductor industry has changed considerably over the course of this author's thirty-five years of involvement in it—in structure, in size, in the stability of the technology, and in the way in which it pursues R&D. In the early days, it was not clear what the mainstream technology would be. Someone would come up with a bi-polar transistor; someone else would propose a thin film device. The direction was finally established with the advent of semiconductor memory in the one, four, sixteen, and sixty-four progression. While it used to be that each of the major players had a fairly important laboratory conducting basic research, much of that has disappeared. Intel, as noted, limits internal basic R&D to what is needed to solve immediate problems.

In its early years, the company looked to Bell Laboratories for basic materials and science related to semiconductor devices; it looked to RCA's Princeton labs for consumer-oriented product ideas; and it sought insights into basic materials problems and metallurgy from the laboratories of General Electric. Over time, Intel found that most of the basic R&D relevant to its needs was being done by companies such as Fairchild and Texas Instruments, which had evolved into the product leaders in semiconductors.

Today, Intel looks to universities for much of the basic research of interest to it. That university chemical and electrical engineering departments have caught up with industry needs is evidenced by the Semiconductor Research Corporation (SRC) established in the early 1980s by the Semiconductor Industry Association (SIA). SRC essentially taxes large numbers of firms and users and then deploys the monies raised to promote university research that has relevance for the industry. Much of the university research conducted prior to the creation of the consortium had concerned some of the more esoteric materials, 3-5 and 2-6 semiconductor compounds, for example, rather than silicon. The consortium's funds have been successful in keeping several major universities engaged in research that is more immediately germane to the industry. Through faculty interest leading to proposals for government support, the industry has probably leveraged more than two to three times the money it has invested in the SRC, some $200 million over a ten-year period.

As the technology and behavior of semiconductor devices has become better understood over the past decade or two, the emphasis of semiconductor research has shifted from science to engineering. Moreover, the extremely rapid evolution of the technology has rendered the industry less dependent on breakthroughs. With the direction of the technology fairly clear, investment has become the paramount concern. The breakthroughs that do occur will be products of exploratory research done in protected, fertile laboratories in which basic ideas and fundamental understanding are developed and passed along to be exploited at the appropriate time by start-up companies.

The large, central research laboratories of the premier semiconductor firms probably have contributed more to the common good than to their corporations. Bell Labs, for example, did contribute much to AT&T, but its greater contribution seems to have been to the economy as a whole. Fairchild's large research organization, particularly in its later years, probably contributed more to the many spin-off companies that exploited the ideas that surfaced within it than it did to its parent company. More recently, Xerox Corporation's Palo Alto Research Center made some tremendous contributions to the community at large, notably in the area of local area networks and the graphical user interface that became the basis of the Macintosh computer. Xerox itself, however, did not benefit nearly as much as it might have from these developments.

Why do spin-offs and the community at large tend to reap so much from large research organizations and the firms that own them so little? In part, because of the nature of the industry. Most semiconductor companies are cross-licensed with other semiconductor companies, thus intellectual property is not easily protected. Furthermore, the companies that own these laboratories have an inherent disadvantage. Because these companies are large, successful, and established, they tend to have difficulty exploiting new ideas. Even an idea recognized to have the potential to be important in a few years is not likely to be accorded sufficient attention and assigned the best people, preconditions for germination. Running with the ideas that big companies can only lope along with has come to be the acknowledged role of the spin-off, or start-up. Note, however, that it is important to distinguish here between exploitation and creation. It is often said that start-ups are better at creating new things. They are not; they are better at *exploiting* them. Successful start-ups almost always begin with an idea that has ripened in the research organization of a large company. Lose the large companies or research organizations of large companies, and start-ups disappear.

The biotechnology industry presents a somewhat different model for sourcing basic research. What Bell Labs was to semiconductors in the early 1960s, the National Institutes of Health are to bioscience today. Large, government-research-funded structures support other industries, notably agriculture and nuclear energy. In setting up what is essentially a central research laboratory, the semiconductor companies that are allied in the Microelectronics and Computer Technology Corporation (MCC) are grasping at such a model.

While some have suggested that the university research machine, unquestionably the best in the world over the long term, should assume a leadership role in creating critical new technology, this simply is not practical in the semiconductor industry. It would require an enormous investment just to equip universities to work with the state of the art of the technology. To remain at the leading edge, an equivalent investment would be required about every three years. A number of smaller systems companies that decided they needed semiconductor technology in-house embarked on this path. They spent $50 million to acquire the technology, $70 million the next year to upgrade it, and $90 million the following year, and they had drawn only the first card.

If it is impossible for universities to afford the equipment required to support work with state of the art semiconductor technology, what is reasonable to expect in the way of relevant research? The process of producing integrated circuits comprises many "chunks." The scientific content of some of these chunks is very important. For example, production of semiconductor structures has shifted from a wet etching process to a plasma etching process, but plasma etching is not well understood, rendering its execution more of an art than a science. As the process does not require integrated manufacturing capability to examine, university researchers could further the understanding of plasma etching using relatively simple equipment. The semiconductor industry could then employ that knowledge to better engineer the process. Many similar prospects for academic research can be identified. Parallel processing is another example of the kinds of research at which universities excel. Cal Tech, which is a leader in parallel processing research, is teaching undergraduates how to program parallel machines instead of single processors. Universities do this sort of thing very well. Finally, much of the software produced by universities is potentially useful to the industry. Some of the design automation software originated by universities can be used almost without modification.

For its part, the semiconductor industry has undertaken a number of initiatives aimed at promoting a more efficient use of its resources. The founding of the Semiconductor Industry Association (SIA) in the 1970s was essentially an expression of concern about the ascendancy of the Japanese semiconductor industry. The association was formed with the idea of presenting to the federal government a unified image of the state of the industry and soliciting its help. Subsequent initiatives included the Semiconductor Research Corporation and SEMATECH, which was formed to address U.S. semiconductor firms' inferior, when compared to their Japanese counterparts, manufacturing capability.

Early in SEMATECH's formation, its founders organized a series of industrywide workshops to identify the technological advances required for U.S. semiconductor and supplier industries to catch up with Japanese industries. The outcome, in March 1988, was a timeline and the specifications for a sequence of technological generations that would lead to parity by 1994—a "road map for semiconductor technology." The timeline specifications required the demonstration of a 0.8 micron technology in SEMATECH's new wafer facility in 1989, with further advances to 0.5 micron technology in 1990, 0.35 micron technology in 1992, and 0.25 micron technology in 1994.

In 1988, Congress established the National Advisory Committee on Semiconductors (NACS) to "devise and promulgate a national semiconductor strategy." The committee, which included leaders from both industry and government, published a series of recommendations for strengthening the nation's semiconductor industry. These included a technological projection, Microtech 2000, that called for the industry to accelerate development by a full generation by the year 2000. In April 1991, the Technology Committee of NACS and the Federal Office of Science and Technology Policy cosponsored a workshop at the Research Triangle in North Carolina. Some seventy representatives from U.S. semiconductor manufacturers, equipment makers, research institutions, universities, material suppliers, and the federal government examined Microtech 2000 in detail. The results of the workshop were widely circulated and created a great deal of discussion in the semiconductor and equipment industries, although the goal of accelerated development was not accepted.

When the NACS activity ended in 1991, the U.S. semiconductor industry was asked to take over Microtech 2000. The SIA formed a technology committee that sponsored the formation of a group to update Microtech 2000 in a workshop held in the fall of 1992. The SIA workshop took a slightly different approach. There were at least two objectives. The first was to coordinate the long-range activities of the SRC and SEMATECH. The second was to provide a road map for fifteen years, pointing out key technology needs and the times at which those technologies would be required to keep the semiconductor industry on the historic productivity curve of a 30 percent reduction in cost per function per year. Unlike the earlier road map, this map published in the spring of 1993 included cost per square centimeter as a benchmark metric against which technology had to perform in the future. Although the map was strictly a U.S. effort, it was made available worldwide at no cost.

Rapid advance in the semiconductor industry prompted an update of the map within two years. The *National Technology Roadmap for Semiconductors* was published in the spring of 1995. The time frame extends to the year 2010, when the minimum feature size is expected to be 0.07 microns. The availability of an authoritative technological map should promote more efficient use of resources, limit overlap, and avoid gaps. Certainly, the experience of producing it has been unique in U.S. industry.

The federal government spends several hundred million dollars per year on semiconductor research. The present technology map is an effort to make the federal government spend more of that money on research that the industry believes to be important. The map is not likely, as some have worried, to lead the industry over a cliff. The directions are clearly those the industry must follow if it is to survive. While there have been concerns that such a map may draw resources from existing work that may be more important, individual investigators seeking funding are likely to use it only as it makes sense and strike out on their own when they suspect they have better ideas. It is, all in all, more important not to dissipate resources through a failure to recognize problems and their relative importance.

Finally, there is the question of where in the world breakthroughs will occur. Will they originate in the United States or offshore? If offshore, who will exploit them? One thing U.S. industry has demonstrated consistently is its inability to exploit developments that originate abroad. This seeming inability should be cause for considerable concern.

PART THREE

RECONCEPTUALIZING RESEARCH AND INNOVATION

CHAPTER 8

Commercializing Technology: Imaginative Understanding of User Needs

Dorothy Leonard-Barton and John L. Doyle

CHAPTER 9

Rethinking the Role of Industrial Research

Mark B. Myers and Richard S. Rosenbloom

CHAPTER

8

COMMERCIALIZING TECHNOLOGY: IMAGINATIVE UNDERSTANDING OF USER NEEDS

Dorothy Leonard-Barton and John L. Doyle

THE STATISTICS ON technology commercialization in the United States are fairly grim. On average, as much as 46 percent of all resources devoted to product development and commercialization are spent on products that are cancelled or fail to yield adequate financial returns (Cooper 1986, 16). A Booz-Allen report (1982) estimates that for every 100 projects that enter development, 63 are cancelled, 25 become commercial successes, and 12 are commercial failures. Even so, some companies boast a success rate of 70 to 80 percent in new product launches. Clearly, then, it is possible to beat the averages through a superior product development process (Cooper 1986, 17). Two criteria dominate the definition of superiority in product development today: speed to market and relevance.

The ever-shortening product life cycles in many industries is one of the major forces driving an increasing focus on improving development practices (von Braun, 1991). Hewlett-Packard has documented these shortened life cycles in its own products as Figures 8-1 and 8-2 demonstrate. In Figure 8-1, each line on the graph represents the sales history over time of all HP products that were launched in the year at which the line originates. Thus, for example, the cohort of products launched in 1979 brought in about $250 million worth of orders in that first year and increased to $600 million in 1980, which was its highest year. Thereafter sales gradually dwindled off. In contrast, the cohorts of products introduced in 1987 and in 1988 produced a steep rise in sales the following year (1988 and 1989) but dropped off dramatically thereafter.

Using their expertise in measuring waveforms, Hewlett-Packard managers translated the data in Figure 8-1 into sales windows for

Figure 8-1.   Product Sales History*

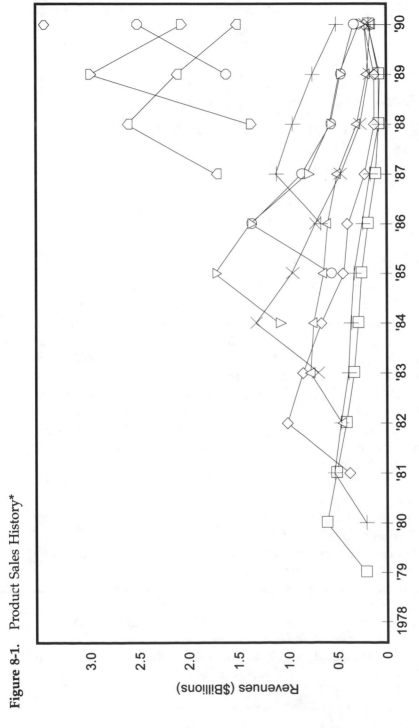

* Each line on the graph represents the sales history over time of all those products launched in the year at which the line originates.
Courtesy of Hewlett-Packard.

Figure 8-2. Sales Windows for Product Cohorts*

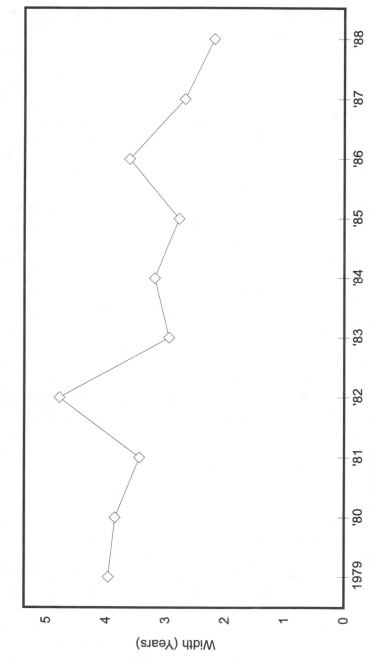

*Each point on the graph indicates the number of years between (1) the year sales of a particular cohort of products first reached one-half its subsequent sales peak; and (2) the year when sales fell to that one-half peak level.

Courtesy of Hewlett-Packard.

Figure 8-2. In Figure 8-2, each point on the graph indicates the number of years that fall between two points: the year that the sales of a particular cohort of products first reached one-half its eventual sales peak; and the year that sales fell to that level after peaking. As the graph in Figure 8-2 indicates, the sales window has been narrowing over time as the curve describing the rise and fall of sales orders for a given cohort of products has become both shorter and steeper.

Shorter product life cycles have led companies to emphasize time as the basis of competition, vying with each other to publicize bringing product to market in record time (Bower and Hout 1988; Dumaine 1989; Stalk 1988). In recounting their successes, however, managers sometimes conveniently overlook the reservoir of knowledge that has been tapped. In other words, the publicized project time lines do not include the time spent developing support tools or conducting laboratory experiments. Abortive prior attempts to build similar products are rarely acknowledged, although these attempts may have yielded critical knowledge. Digital Equipment, for example, brought its first engineering workstation, the 3100, to market in an unheard-of eight months. Besides departing from tradition to buy a MIPS company semiconductor chip, DEC utilized UNIX-based computer-aided design tools that were available only in its research laboratories, and mined graphical interface designs that had been developed for prior, cancelled projects. Developers candidly admitted they could not have made their frenetic dash to the market without the information garnered from these prior projects (Bowen et al. 1994). Kodak was able to bring out its Funsaver, single-use camera in only nine months when it was challenged by Fuji principally because an engineer had been working on the camera for two years with no official encouragement or even sanction (Bowen et al. 1994).

As these examples suggest, speed to market often depends upon the development team's ability to pull unutilized but technically developed concepts, designs, and tools from the knowledge warehouses of the corporations. Such reservoirs of knowledge, which are built up over time and shaped by the norms and values of the corporation, the skills of its personnel, and the particular incentive systems in place, can constitute unique technological competencies (Leonard-Barton 1992a). The notion that companies possess distinctive competencies that can be systematically deployed for competitive advantage has existed for decades (Rumelt 1986; Hayes 1985; Hitt and Ireland 1985). It has received renewed attention lately because of disenchantment with strategies based on accumulating unrelated businesses, because of the recognition that Japanese competitors have exploited their capabilities shrewdly (Prahalad and Hamel 1990), and because successful new products often build on those competencies (Maidique and Zirger 1985).

At least superficially, this renewed interest in the accumulation of unique knowledge seems to conflict with the move in the corporate world away from investments in central laboratories. The principal argument cited for this move has been the inability of corporations to turn knowledge into practical applications and profit, which has led corporate laboratories to feel increasing pressure to demonstrate their utility. Adding to this pressure is the growing realization by managers that the distinctive competencies of a firm depend upon knowledge diffused throughout the organization, not just in the research laboratory. Some managers have even denounced the idea of a special research facility. Chaparral Steel, one of the most successful U.S. minimills, is extremely innovative and has patented a number of inventions, yet it does not separate research and development from production (Leonard-Barton 1992b). The company's CEO, an outspoken former R&D director with a Ph.D. in metallurgy from MIT, maintains that research laboratories are idea graveyards "not because there are no good ideas there, but because the good ideas are dying there all the time" (Kantrow 1986, 99).

For a manager of research in the 1990s, then, the challenge is to produce relevant, potentially profitable knowledge faster than ever before. To meet this challenge successfully the research organization must understand user needs, which is the focus of this chapter. We first briefly examine the evidence that links imaginative understanding of user needs to successful technology commercialization. We next explore what it means to understand user needs and the management dilemmas inherent in that process. The second section of the chapter proposes a typology of technology commercialization situations and describes mechanisms for understanding user needs when products are developed in advance of the current market. In the final section, Hewlett-Packard laboratories are used as an example of a deliberate balance struck between the huge, highly autonomous corporate laboratories built by some companies and the diffused pockets of research spread among business units favored by other companies. Hewlett-Packard's success with such innovations as the deskjet printer has grown out of the company's ability to create a convergence of technology trajectories and business and market strategies.

UNDERSTANDING USER NEEDS

Systematic research on product success and failure has shown that failure begins with an inadequate product definition process. A study by Edith Wilson (1990) of nineteen successful and unsuccessful projects in nine of Hewlett-Packard's fourteen business groups revealed that the unsuccessful ones suffered from failure to understand user needs. Although the teams behind these projects often had spent months and millions of dollars in a vain attempt to answer some fundamental questions about their customers, they could not translate user needs into product. They had problems in identifying the target user, the actual buyer who controlled the financial decision, and the other stakeholders who affected the buying decision or they were unable to determine exactly what problems had to be solved to satisfy each link in the set of customers from the factory to the end-users.

This major observation from Wilson's study (and the SAPPHO project that inspired it, see Rothwell et al. work [1974]) has been made by a number of other new product development studies as well. In fact, problems with product definition are endemic to development (Gupta and Wilemon 1990; Cooper 1986; Rothwell et al. 1974). Hopkins and Bailey (1971) found inadequate market analysis most commonly cited as a cause of new product failure, and in a number of studies conducted over the past two decades, Robert Cooper and his colleagues have identified inadequate attention to market as a primary factor leading to failure (Cooper 1975; Calantone and Cooper 1979; Cooper and Kleinschmidt 1986; Cooper 1986). In a study of 235 new product development projects, Souder (1987) also cited clarity of problem definition, which he equates with clarity of understanding user requirements, as a primary condition for technical success.

While there is general agreement that an understanding of user needs is one of the key factors for commercialization success, there is little agreement on how that understanding may be achieved. In their study of 252 product development projects in 123 firms, Cooper and Kleinschmidt (1986) found that preliminary market assessments were conducted in the successful product development projects. However, formal market studies, done in only one-quarter of the projects, were usually rated as "poorly handled" (p. 78). Moreover, the formal market studies tended to take the form of reactive competitive comparisons in over one-fourth of those projects. There were almost no tests studying customer reactions to a proposed new product in concept form, and less than one-fifth of the project teams studied what customers actually wanted or needed before generating product specifications. As there was at least as much detailed market research done for failed projects as for successful ones, simply increasing the emphasis on market research does not appear to lead to a better understanding of user needs and a higher probability of product success.

A survey of these studies on the importance of understanding user needs indicates that there are at least three major barriers to introducing market-derived information successfully into a new product development project: corporate core rigidities, the tyranny of the current market, and user myopia.

Core Rigidities

As suggested above, the core competencies of a firm often aid the commercialization of technologies. However, the very same core technical capabilities that have made a company great can constitute core rigidities and hinder new product development in subtle ways (Leonard-Barton 1992a, 1995). New product ideas built on familiar technologies, using traditional, comfortable sources of information are more easily commercialized. In fact, synergy with the firm's capabilities (technological resources and skills) has been identified as one of the factors "fundamental" to new product success (Cooper and Kleinschmidt 1986, 1990; Souder 1987). New product ideas built on unfamiliar technologies are often seen as "illegitimate" (Dougherty and Heller 1994). Thus, these products see a more difficult birthing, and so the ventures are often isolated from the rest of the organization (Kanter 1988; Burgelman 1983). Well-entrenched routines and ingrained corporate culture favor certain technologies and information sources. In "technology-driven" companies, information about user needs often goes unheeded unless it comes from a source with status in the organization. The same characteristics and practices that constantly reinforce the ability of engineers to influence product design simultaneously undermine the ability of marketing (or manufacturing) people on the team to be heard.

Companies such as Kodak, Digital Equipment, and Hewlett-Packard owed their original success to technological innovation, and the primary sources of that innovation were the technologists. In these companies, the product design intuition of founders and other technical "gurus" early in corporate history is legendary. Some researchers in Hewlett-Packard Laboratories can still recall their skepticism that the handheld calculator championed by Hewlett could ever function better than their slide rules; Gordon Bell at Digital Equipment is credited with the "one architecture" strategy that made the VAX line of minicomputers so popular. Over time, these companies developed a strong technological capability comprised of skilled engineers and proprietary physical equipment (simulation systems, process equipment) that embodied years of accumulated, specialized knowledge. Hiring practices, incentive systems, and the values and norms of the company strengthened that capability by enabling the corporation to attract and hold the best technical minds in certain fields. For instance, skilled chemical engineers migrated to Kodak, where they could aspire to the pinnacle of success represented by the top 5 percent of engineers who became film designers. Electrical and mechanical engineers headed for Digital or Hewlett-Packard. In contrast, software engineers entered any of these companies at a disadvantage, knowing their profession was not highly regarded, and marketing people entered at their own peril. Their status and ability to influence key product design decisions were subordinate to those of researchers and design engineers. Formal marketing skills were not part of the original distinctive competencies of these firms. Thus, the skills they brought to the firm often were not fully employed or were employed late in a project, resulting in delays. For example, during the design of the Deskjet printer at Hewlett-Packard, marketers tested early prototypes in shopping malls to determine user response. They returned with a list of twenty-one changes they believed essential to the success of the product. The engineers accepted five. Unwilling to give up, the marketers persuaded the engineers to join them in the mall tests. After hearing from potential users the same feedback that they had previously rejected, the product designers incorporated the other sixteen requested changes. At DEC, early in the development of a local area network switch, marketing personnel suggested a number of features that engineers rejected as unnecessary. One month before expected release of the product, a respected senior engineer visited several customers and discovered the rejected features were essential. The schedule was slipped several months to allow the design to be retrofitted. In both these cases, information about user needs was available, but the marketing people lacked the experience and status to influence product design. As a consequence, the product definition was altered late in the development process and the products reached the market later than necessary (Leonard-Barton 1992a).

The Tyranny of Current Markets

On the other hand, listening too closely to current markets can also constitute a barrier to commercializing technology. Responding to a flood of marketplace demands for improvement along current product performance curves can leave too few resources to assess the possibility that new technologies are altering those curves. Thus, traditional sources of market information and influence on new product development can also constitute a core rigidity for a company. Christensen (1992) researched four architectural transitions represented by the reduction in computer disk drive diameter from 14 to 8, 5.25, 3.5, and 2.5 inches. Each reduction entailed not only "shrinking" individual components, but re-architecting the relationships of components within the system. Christensen found that existing firms tended not to introduce these changes because their customers were not interested in the smaller disks.

> The sluggishness or failure of established disk drive manufacturers faced with architectural change seems rooted . . . in the inability of their *marketing and administrative* organizations to find customers who valued the attributes of the new-architecture drives. . . . not . . . because their architecture-related *engineering* knowledge was rendered obsolete. . . . Rather, the changes in product architecture seem to have rendered obsolete the established firms' knowledge of *market.* (p. 114–15)

In the few cases in which the new technological architectures did appeal to a firm's current customers, existing firms, rather than entrant firms, dominated. For example, Conner Corporation was able to make a "very smooth transition into 2.5-inch drives [from 3.5 inch]" because the smaller disks appealed to current customers (p. 150). The situation studied by Christensen exemplifies a common problem in many companies: "Outbound" marketing efforts (selling) tends to supersede "inbound" marketing efforts (market research and development). The impact on the bottom line of the outbound effort is much more visible and immediate than that of inbound, and managers rationally emphasize those activities for which they are most directly rewarded. Moreover, both in academia and within corporations, more training is devoted to outbound promotion and selling activities than to gathering and translating market information into actionable commercialization steps.

Users' Natural Myopia

Finally, as the example of the computer disk industry suggests, the greatest challenge to understanding user needs lies in the need to select the right users as informants and to recognize when their suggestions may limit product design. Users are often myopic in a number of logical and natural ways. First, they see the potential to apply technology within their own bounded context and naturally will influence the design of the new product or process to meet their needs within that particular environment. Software designers constantly face this problem. For example, a vendor who was designing a purchasing module for the manufacturing resource planning system to order and monitor purchased parts for a multiplant corporation solicited the help of users in a certain plant in the Northeast, assuming that the plant was representative of all corporate plants. The flaw in this assumption was spotted only after managers of shipping docks in a dozen other plants angrily reported being so buried every four months with thousands of long-lead time purchased parts that they had to hire extra help. At the plant that had served as the model, only 15 percent of the parts ordered were long-lead time items; therefore a regularly scheduled delivery point of every four months was adequate. For all the other plants in the system, however, the parts represented over 40 percent of their purchased parts. These plants required that the flow be spaced out over time (Leonard-Barton 1989).

Second, users are not all equally proximate to the latest trends in usage patterns. Designers of a system to monitor the progress of work-in-process inventory within a factory made sure their software could accommodate the variety and volume of components moving down the line. Within weeks of installation, however, their system was obsolete because the factory moved to a just-in-time system for which any build-up of inventory was an anathema (Leonard-Barton and Sinha 1990).

Finally, users cannot see their world through the eyes of the technologist and therefore cannot know what solutions, functions, enhanced features, or capabilities a technology may offer. Technology always offers more possibilities than can be recognized and commercialized. Establishing paternity through DNA tests of blood samples is not the most obvious application of the discovery of DNA, and holographs on greeting cards are not the most profound application of holograms. Therefore, while technologists cannot ignore user needs, they often cannot simply ask users what they want. Instead, developers need a whole range of identification, listening, and translation skills to translate user needs into commercialization opportunities.

A TYPOLOGY OF COMMERCIALIZATION

Technology commercialization situations range from situations in which technological potential aligns with current markets (far left in Figure 8-3) to situations that require the creation of new markets, sometimes even whole new businesses with new infrastructures, standards, and procedures (far right in Figure 8-3). When technological potential aligns with current markets, product definition and development follow quite well-known paths. The risk and uncertainty inherent in product design are low and so is the difficulty in communicating the product concept to others (see Figure 8-4). In contrast, when new markets or even businesses must be created, commercialization is realized only through a combination of trial, perseverance, and serendipity. Clearly, in these cases, risk and the difficulty in communicating a product concept are high.

An Improved Solution for a Known Need

Competition or explicit customer demands often drive technological improvements along known performance parameters for current products. In these cases, with or without extensive market research, developers often know that lower costs, more features, or better quality are likely to win in the marketplace. Hewlett-Packard's Signal Analysis Division, which produces spectrum analysis devices for testing and analyzing radio-frequency and microwave signals, had traditionally competed on product quality and performance rather than price. In the mid-1980s, however, the commercial market expanded, particularly at the low end where Japanese offerings threatened Hewlett-Packard products. Although a low-cost spectrum analyzer existed in the laboratory, there was little interest in commercializing it until an R&D manager returned from a plant visit with a customer in Italy who pointed out that the Japanese had produced a low-cost product with features comparable to Hewlett-Packard's high-priced spectrum analyzer. Intent on bringing the Hornet, as the analyzer was called, to market within eighteen months, the team nonetheless took time to conduct an unprecedented price study. Moreover, marketing personnel accompanied engineers on customer visits to assess user needs, al-

Figure 8-3.   A Typology of Technology Commercialization Situations

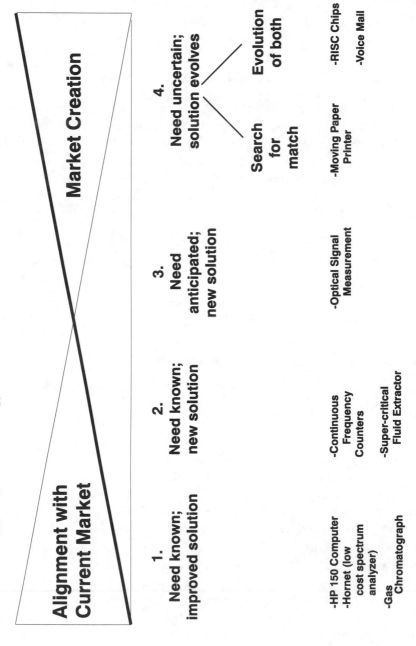

Figure 8-4.   Challenges of Designing in Advance of the Market

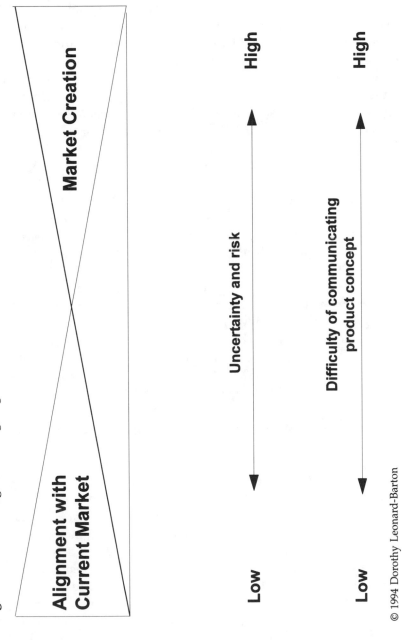

though R&D maintained responsibility for product definition. The Hornet met both cost and schedule goals and was a marketplace success.

Less successful was the Hewlett-Packard 150, an early attempt to produce a personal computer. The day after the HP120, a terminal for use with the HP3000 minicomputer, was introduced, IBM introduced its first PC. Hewlett-Packard responded by changing the project charter for the HP150, a follow-on to the 120. Under the changed charter, the HP150 was now expected to have enough computing power to stand alone and to be capable of supporting MS-DOS, the operation system for the IBM PC. However, the marketing plan for the HP150 as a terminal for the HP3000 was not altered to reflect the very different set of users now targeted. Therefore, the product development team continued to optimize on the performance characteristics suitable for the customers originally envisioned. As a terminal, the HP150 was quite successful; as a personal computer, it was never profitable (see Bowen et al. 1994).

In both of these examples, developers established user needs in reactive mode. That is, the competition defined the meaningful parameters on which the development team would attempt to achieve parity if not superiority. There were clear benchmarks against which to design. The challenge lay in identifying the right set of users to interview. Once correctly identified, those users could readily communicate their needs, which were used to guide design trade-offs.

Because they know current cost or functionality barriers to usage, developers are aware of a current need in the marketplace even in the absence of direct competition or customer demand. In these cases, developers proactively decide to "delight" their customers with leaps in performance that no competitors have attempted and no users have requested. The challenges of pushing beyond current technical barriers can be very significant. In the early 1980s, Hewlett-Packard was market leader in gas chromatographs, which was an old, mature market. Around 1983, managers in the analytical business decided on a bold target: a chromatograph with one-third of the components of the current model, three times the quality, and one-third of the current price. Such a product was not just a logical enhancement of the current chromatograph. The design required a tremendous leap in performance. In fact, as James Serum, group R&D manager for Hewlett-Packard's analytic group, noted in an interview in November 1992, it "wasn't even on the same price/performance curve," yet the developers "knew in their hearts" that the customers would want it. Customer forums and site visits to observe user practices helped developers to identify critical features. As expected, the product was highly success-ful not only in its traditional market but also, because of its lower cost, in new application areas.

New Solution to a Known Need

Sometimes, users have a need for which they are incapable of imagin-ing a solution because they do not know the technological potential exists. For example, Hewlett-Packard researchers developed a super-critical fluid extractor for the environmental testing market, because they knew that current techniques for analyzing toxic residues in soils involved dangerous solvents and such unpleasant, labor-intensive tech-niques as boiling the soil and reconcentrating it to analyze the residue. Because the researchers were aware of the laboratory experimentation conducted under super-critical conditions, they could imagine an ap-plication in which carbon dioxide in a super-critical, near-liquid state is passed through soil to extract pesticides. Once the carbon dioxide is released from the pressure that renders it super-critical, it returns to an easily vented gaseous state, leaving behind the extracted organic sam-ple. After conducting market research to establish uses of the technique, the team built a prototype and had lead customers run their samples through it. The extractor connects to other Hewlett-Packard analytical instruments, and its use improves productivity significantly. Currently used in Environmental Protection Agency laboratories, the extractor is not as high volume and profitable as the gas chromatograph, but it has solved an array of real customer problems.

The Hewlett-Packard continuous frequency counter introduced in 1986 originated in a somewhat similar situation. Traditional frequency counters for monitoring various types of waves gradually became outdated as transmitters generating frequencies increased in accuracy. One of the company's engineers envisioned a counter analogous to an oscilloscope that could read frequencies continuously as a series of digits that could then be plotted on an X/Y display and could track drift in frequency signals. However, even this analogy did not help when marketing described the new concept to prospective users, who were markedly unenthusiastic. Convinced that a current (if unrecog-nized) need did exist, the engineers constructed a functional prototype, which they persuaded marketing to take into the field. Somewhat to everyone's amazement, the prototype was seized on by users with such enthusiasm that marketing sometimes had a hard time retrieving the models. Users saw immediate applications, including many that the engineers had not anticipated. One customer wanted to hook the counter up to his radar system to check its functioning. The product became a great commercial success, representing approximately 15 percent of the division's sales. Even more, it reinvigorated a product line that had been in decline.

A New Solution to an Anticipated Need

If meeting unrecognized but current needs is difficult, peering into the future to identify the as yet unarticulated future needs of a given market is more difficult. By extrapolating societal, technological, environmental, economic, or political trends, developers attempt to foresee what users will need when those trends mature. To someone who understands both industry and societal trends, the needs themselves may be fairly obvious. Timing, however, often is extremely unclear. When will there be enough users or complementary technology or adequate infrastructure to justify development? Moreover, trends interact. For instance, society's demand for faster, more proximate means of communication interacts with the technological trends that are driving computers to become commodities and communications to become wireless.

Many companies are now tracking environmental trends. Legislation in Europe mandating that companies accept their products for recycling may foretell similar moves in U.S. markets. At present, however, the supply of recycled materials outstrips both our capacity to process them and the demand for their reuse. Industry is thus caught in a Catch-22: Companies cannot recycle materials economically until they can process them on a large scale, and consumers do not want to buy recycled goods because they are too expensive. To create a market for recycled materials, then, companies must create collection processes, create new process technologies to accommodate recycled materials, and stimulate demand for products designed from recycled materials.

An Evolving Solution to an Uncertain Need

At times, technologists run far ahead of consumers by developing an application for which they initially identify the wrong market. When Hewlett-Packard researchers first recognized their ability to develop a printer that moved a pen across a plotter and the paper underneath the pen in the other direction, they were enthusiastic. The opportunity for small-scale, high-resolution printers seemed obvious to them, and they were dismayed to find their enthusiasm was not shared by the division producing large printers. Only when they were able to engage the interest of the medical division in using the printer for plotting electrocardiographs was a market identified. That technology eventually repaid the relatively modest costs of its development many times over. Cases like these are often disparagingly called "technology push" in recognition of the fact that the technical possibility preceded any known user need. Laboratories and the basements of home inventors are full of failed solutions to unknown problems. Still, the negative connotation of the phrase is misleading in two ways. First, there are many products on the market for which no user felt or expressed a need, and also which embody no technology. The notorious "pet rock" sold in the 1970s may not exemplify "technology push," but it certainly exemplifies "sellers push." Second, there are many extremely well-known inventions for which there was initially no user demand, but that many people today insist they need, such as xerography or post-it pads. Sometimes need and solution evolve together. Two widely used technologies that started life in quite different forms at IBM were shaped by trial and error and through the brutal help of internal corporate selection processes and the marketplace. During an interview in November 1992, Joel Birnbaum, vice president and director of Hewlett-Packard Laboratories, noted that the voice-mail systems that are so ubiquitous today originated at IBM when a remote dictation system was designed to enable traveling managers to relay their correspondence over telephone lines to a pool of specially skilled typists in a manuscript center. The so-called Advanced Dictation System was not used as expected. Both managers and documentation preparation people disliked it, but people started sending messages back to their secretaries using the system, which was retitled the Audio Distribution System. In the early 1970s, this system became the Speech Filing System—the prototype of current voice mail.

Birnbaum also noted the reduced instruction set chip (RISC) so widely used today started life in a form very different from today's form. In the early 1970s, when the eventual intersection of telephones and computers was foreseen, researchers at IBM saw a need to apply digital computing technology to digital switching for telephones. Since such switches have extremely long lives, the computer software that underlay their design needed to be structured to evolve and the hardware needed to be scalable. The computer architecture invented to meet this need was not used for this because of IBM business decisions.

Instead, it was generalized and extended for less specialized uses. In the mid-1970s, it became the basis for what is known today as RISC architecture.

At the right side of the spectrum pictured in Figure 8-3, there is a strong element of serendipity. Technological potential and market need have to coincide in both time and place for the necessary synergy to occur. However, research managers can encourage that serendipity by building their intuition about the future.

TOOLS AND MECHANISMS FOR UNDERSTANDING USER NEEDS

When the product is well aligned with current markets, traditional market research can help track customers' needs for the needs are clear, current, and communicable. At the opposite end of the spectrum, when the uncertainty of investing in technology for an unknown market is so great, market intuition and serendipity play a large role and it is not clear the process can be as systematized and structured. In between these two points, however, are known needs and anticipated needs. These situations are candidates for considerably more managerial attention for here lie largely unexplored opportunities to improve the process of technology commercialization through empathic design (Leonard-Barton 1995).

Empathic design, as the term implies, means understanding user needs through *empathy* with the user world rather than through user articulation of needs. The significance of empathic design is lower at either extreme of technology commercialization (see Figure 8-5), but still important to understand. Even when the product that is being designed is aligned with current markets and the users are well known, there are desirable product attributes that the user is often unable to articulate or even imagine. Often, users are totally unaware of their own psychological and cultural responses to symbols and forms. They are unable to describe what they want until they see it. Therefore, so-called intangibles such as shape, appearance, and feel are usually determined by expert designers through empathic design. At the market creation end of the spectrum, the ability of developers to imagine a future user environment in which their technology could function clearly requires some degree of empathic design. Using sketches or models by industrial designers to embody imagination is particularly useful in envisioning potential futures as no real users have

Figure 8-5. Tools and Mechanisms for Understanding User Needs

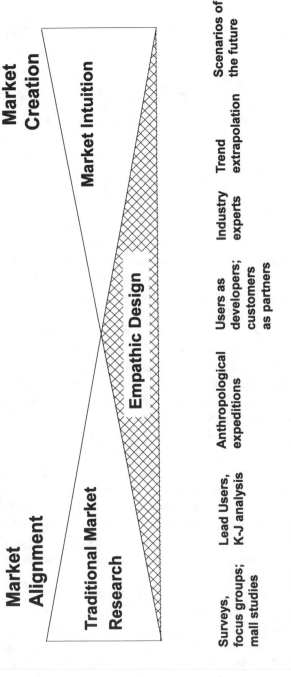

© 1994 Dorothy Leonard-Barton

been reliably identified (Leonard-Barton 1992c). However, empathic design is most powerful when developers are proposing new solutions, to known or anticipated user needs. In these situations, as product designers develop a deep understanding of the current user environment, they can extrapolate about the way in which that environment may evolve and imagine the future needs their technology can satisfy.

Tools that help to explore user reactions to products at the alignment end of the spectrum are widely researched and taught. Marketing departments in the western hemisphere are experts at conducting surveys and focus group interviews. The mall studies mentioned in the description of the changes made to the Deskjet printer exemplify more ambitious but widely used techniques. Less widely employed are interviews with so-called lead users who are already at the cutting edge of current products (von Hippel 1983). By anticipating their own immediate needs, these users can foretell the more distant needs of less advanced users and guide the design of radically improved solutions. Other techniques are most helpful when users can identify a product concept, but important dimensions of the concept are unclear. For example, the KJ analysis, which was first developed by Jiro Kawakita, a noted Japanese anthropologist, is a highly structured process for gathering and analyzing qualitative data. It has been applied to quality improvement programs in Japan and, with the help of Professor Shoji Shiba of Tsukuba University, in the United States (Shiba et al. 1991). It has also been applied to "concept engineering" at the Center for Quality Management, Cambridge, Massachusetts.[1] Another example of systematic need elicitation is the proprietary Value Matrix, which is used by the Design Consortium to uncover the often latent desires of a wide spectrum of user/stakeholders in client firms.

The most powerful aid to empathic design is an anthropological expedition of some kind (Leonard-Barton 1995). Technologists immerse themselves in the user world, much as anthropologists do when they inhabit native villages. Designers or developers with a thorough knowledge of technological potential live in the user environment and absorb enough understanding of that environment to empathize with user needs. They see how users cope with unnecessarily inconvenient, uncomfortable, inefficient, or inaccurate tools and then consider how to solve the unspoken problems. Products designed through this process present users with functionality, ease of use, and other benefits that they would not have thought to ask for. One of the most famous designers in U.S. history, Henry Dreyfuss, used to require his designers to live with whatever tool they were designing. They rode corn pickers and prowled factories. Product developers in Hewlett-Packard's medical division spend time in intensive care units and hospital clinics. It was on such a visit recently that a product developer noticed that nurses, in the course of their duties, inadvertently blocked a surgeon's view of the television screen which was used to guide the intricate work. To solve this problem, the product designer conceived of a tiny screen mounted on a surgeon's helmet that would keep the image directly and constantly in view. In a different type of anthropological expedition, Xerox Corporation has employed trained anthropologists and other behavioralists to investigate exactly how people interact with sophisticated copier machines and how they report findings to the corporation as a whole. These explorations resulted in knowledge about users' assumptions and mental models that was pertinent to machine design but had never been systematically gathered (Brown and Newman 1985).

Of course, one obvious way to commercialize technology is for users to become developers, that is, for those who have experienced a problem to apply technology in its solution. Von Hippel (1988) has documented hundreds of cases in which users were the primary inventors of a new tool or process. In order to bring user-innovation in-house, companies hire customers or individuals who have extensive experience in the user world. At Hewlett-Packard, for example, the product line manager for mass spectrometry and infrared spectroscopy systems has a Ph.D. in organic chemistry with a specialty in mass spectrometry. As Dr. Serum noted in his interview in November 1992, "Most of our senior managers have been born and raised in the [analytical chemistry] business . . . when we speak with customers, the full implications of every word are immediately relevant to us; . . . we can feel the pulse of the customer."

Rather than hiring customers to represent the user world, some companies have pursued a policy of commercializing their basic technologies through partnerships with customers. ALZA Corporation is noted for its nontraditional drug-delivery systems that deliver drugs into a patient's bloodstream over time at a continuous rate. The company partners with specific customers to design customized delivery systems for a particular drug. For example, Janssen Pharmaceutica sells transdermal patches for Duragesic, a painkiller for cancer patients, for which ALZA tailors its membranes. ALZA also partnered with Pfizer to deliver medicine for the treatment of angina that may be taken just once, rather than three times, a day.

Industry experts can also offer a lens into the user world. Allegheny Ludlum, a highly profitable specialty steel producer, set up a market development group almost thirty years ago. While the mandate for group members has changed somewhat over the years, their basic task is to entrench themselves in the personal networks of the customer base they serve. Each member of the group is a walking compendium of intense technical knowledge about certain alloys and their application; personal contacts among industry experts and customer companies; and deep knowledge about standard setting and regulation in their industry.

When the user world lies in the future, developers need to extrapolate current trends and anticipate a world that none can predict confidently. While technology commercialization in this situation usually depends upon the market intuition of well-informed "gurus," there are some techniques for formal scenario-construction (Schwartz 1991). The intent of these scenarios is less to predict exactly a future state than to stimulate consideration of unobvious futures, that is, to divorce thought from a straight, unwavering trend line.

All of these tools and mechanisms have limits and costs. The inability of even very sophisticated market research to predict actual user behavior on occasion is well known. After all, attitudes do not predict behavior and people are often unable to articulate some of their aesthetic and functional needs. Anthropological expeditions are time-consuming and require special skills. Experts hired from the user world can become so enmeshed in daily business activities that they lose touch with the cutting edge.[2] Indeed, the major limitation to creating industry experts is that their expertise is held in their heads. Such tacit knowledge is not easily codified or transferred, and it is difficult to evaluate financially because commercialization ideas planted in one year may not yield sales for five or ten. There is also the danger that industry experts will become too narrowly focused and adopt the same myopic view of the world that their users have. Customer partners, of course, engender similar risks in that they can direct technology commercialization into narrow, self-serving niches. Finally, market scenarios are only intended to provoke possibilities, not to predict the future with any assurance. Therefore, corporations are well advised to pursue a wide range of these techniques to understand user needs. The wider the range available to guide technology commercialization, the more opportunity there is to apply the most appropriate technique to a given situation.

ONE MODEL OF THE FUTURE: HEWLETT-PACKARD LABORATORIES

Hewlett-Packard has an enviable record for the continuous production of new products. Many of these products embody technology that originated in Hewlett-Packard Laboratories. For example, the company's hugely successful Deskjet line of deskjet printers challenged dot-matrix printers on the basis of inkjet technology developed in HP Laboratories. By 1991, HP held 47 percent of the world market in this product category (Finlay 1991). A partial list of other products and business areas that are new to Hewlett-Packard but originated in HP Laboratories is given below. Asterisked items are new business areas for the company.

Instruments
- Surveying Distance Meter*
- Laser Interferometer*
- Cesium Beam Standard*
- Rubidium Frequency Standard*
- Smart Oscilloscope
- Lightwave Instrument Product Line

Medical
- Ultrasonic Imaging Systems*
- Portable Arrhythmia monitor
- Blood Pressure Transducer
- Electronic Clinical Flowsheet

Analytical
- Chemical Experts for IR Spectra
- Capillary Electrophoresis
- UV/Visible Spectrometer*

Computational
- Computer (2116)*
- HP PA-RISC (Precision Architecture)
- Programmable Desktop Calculator*
- Handheld Calculator*
- Moving Paper Plotter
- Laser Line Printer
- ThinkJet Printer

Components
- GaAs Microwave Devices
- Light Emitting Diodes
- Power MOS*
- Shaft Encoders*
- Surface Wave Devices
- Precision Quartz Crystals
- GaAs Integrated Circuits*

Given such a stream of successes in the marketplace, the company's laboratories deserve a close look. On the surface, the central laboratory's mission looks no different from the mission of IBM's laboratories or any other large industrial laboratory: to achieve leadership in international science and technology so as to contribute to the success and profitability of the company. Hewlett-Packard Laboratories is responsible for looking three to five years out, although occasionally a laboratory technology has found its way into a commercial product within a year. There are four major centers. Two laboratories—computer research and measurement—are located in Palo Alto, California. The Bristol laboratories in England focus on multimedia and personal appliances, and a small laboratory (approximately thirty-five people) in Japan focuses on physics and semiconductor research. While the central laboratory's budget (about $156 million for 1994) is under its own control, each laboratory has a designated set of customers in various product divisions that depend upon that laboratory's technological specialty and help set research priorities through constant attention. Approximately ten program reviews a year that are organized around the business/technology intersection (e.g., mass storage, medical systems, imaging) combine with informal interactions between central and divisional laboratories to constitute a system of checks and balances on the central lab's projects. Irrelevant research is likely to be short-lived.

HP Laboratories actually represents a relatively small percent (about 8 percent) of the total R&D budget. The vast majority of research funds goes to supporting the approximately 15,000 scientists and engineers who work in division R&D laboratories. Approximately sixty divisions are clustered in groups, which are combined into sectors. Processors, for example, are part of the workstations group, which is part of the computer systems sector. Each division has total responsibility for the product line it produces, including manufacturing and marketing, and divisional laboratories are very near-term focused.

Traditionally, the company has not competed on marketing or on low cost. Instead, Hewlett-Packard prides itself on its history of innovation. The company was founded by, and is run by, engineers. Managers are steeped in technology and fascinated by it. As Joel Birnbaum noted in his interview, "In a management meeting, if someone brings in a nifty gadget, everyone wants to play with it." Thus, industrial research tends to be engineering-driven rather than science-driven. The tight coupling between business and research is deeply rooted in the culture of the company. When Joel Birnbaum first arrived at Hewlett-Packard, he fully expected to find the transformation of technology into product to be the same problem that it is for many large corporations with large central laboratories. To his surprise, technology transfer was not a critical issue. While certainly technology commercialization is not automatic, or even easy, there are well-established channels for information flow back and forth between the central laboratory and divisions. People in the divisions were interested in what was going on at the central lab, and people at the central lab felt responsible for doing relevant work. The sometimes unspoken but always prevalent expectation was that the laboratories are the source of future businesses. "People at HP truly believe that the way to build better products is to turn up the heat on R&D," Birnbaum observed during his interview. This belief is rooted in the company's history. When the analytic instruments business was under strong market pressure, the company did not sell off the business but rather invested an additional 10 percent in R&D and beat the competition. More recently, Hewlett-Packard Laboratories gambled on photonics by assigning 50 people to this new business and won a leading position in photonics manufacture in the world.

Therefore, at a time when other central laboratories are shrinking or disappearing altogether, Hewlett-Packard is investing increasing amounts in its central laboratory. Birnbaum has noted that the budget for 1994 was up 12 percent from 1993. Even more surprising is the increase in the percent of central laboratory budget devoted to fundamental research rather than applied research (for example, in biotechnology, research on understanding the chemistry underlying genetic engineering as opposed to research on developing a gene-sequencing instrument). While the company's labs will never be considered a likely home for Nobel Prize winners, recent hires have included a higher percentage of theory-driven researchers, whose presence is expected to help Hewlett-Packard avoid missing opportunities at technological cutting edges. The labs are also breaking with the past by increasing the percentage of research monies funding outside work. With increasing pressures on time-to-market, the laboratories find that precompetitive research with university and national laboratory researchers can be highly leveraged.

The theme for the 1990s is to invest in core technologies, including those that cut across businesses. Lewis Platt, Hewlett-Packard's CEO, appears to share Birnbaum's enthusiasm for MC^2: the blending of three leading HP technologies: Measurement (the company's traditional stronghold in test and measurement instruments), Computation (a broad and competitive family of computers), and Communication (computer networking) (Hof 1993). The company should be well positioned to ride the wave of innovation stimulated by new applications for information technology. In 1993, Platt drew together a group of top technologists and marketers from all over the company to constitute an MC^2 council. Their mandate was to narrow to a few dozens the hundreds of possible new markets in the intersection of computers, communication, and networks. From these dozens, a scant three or four will be seriously targeted.

In former times, as much as 85 to 90 percent of the central laboratory activities were focused on supporting current businesses. Today, one-third of the researchers are devoted to seeking new businesses, and many of these work with external consultants to construct forecasts of future markets. Birnbaum predicts: "HP's going to be an almost totally different company 10 years from now" (Hof 1993).

CONCLUSIONS

Not all new products and processes are alike. They may arise either through a process of invention or innovation.[3] While innovation has the characteristics of "being both demandable and predictable within a finite time" (Doyle 1985), invention does not. If one represents the life of a particular technology through the useful oversimplification of an S-shaped curve (see Figure 8-6), it is clear that products and processes arising through invention have potentially a much longer life span than those arising through the more derivative process of innovation (see Foster 1983). Therefore, inventive products are desirable, albeit not to the exclusion of innovative products. However, understanding user needs is absolutely critical to successful technology commercialization of both invention and innovation. Since innovative products and processes àre derivative, users can help share their design. Extensive tools and methods exist for eliciting user needs—even

Figure 8-6. The S-Curve of Technology Maturation

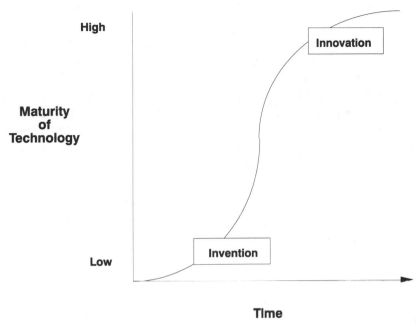

uncommunicated ones—for such products. Unfortunately, usefully channeling user input into the design of inventive products and processes is much more difficult, and there are far fewer tools and methods to apply to the process.

Our current ability to understand imaginatively user needs, i.e., to commercialize technology through empathic design and market intuition, is constrained by the limited number of people capable of this art in any organization. It would seem that the next managerial frontier for technology commercialization is the exploration of methods for understanding current uncommunicated, unrecognized user needs and unarticulated, unobvious future needs. If one primary method for creating this understanding is to harness the creative imagination of individual "gurus," then we need to consider how to identify such individuals and develop their skills to the highest degree possible. Another challenge lies in figuring out how to design anthropological expeditions into user territory (Leonard-Barton 1995). The final challenge lies in developing a better understanding of the limits and possibilities of market scenarios.

A great deal depends upon the vision of the people who invest in the future and upon how far into that future their vision takes them. It is far too soon to say whether Birnbaum's vision of MC^2 will provide Hewlett-Packard with the necessary guidance for a stream of products as successful as those that have emerged from its laboratories in the past few years. However, such a vision certainly enhances the probabilities that the company will be able to recognize and capture profitably the synergies possible across the company's diverse technologies.

In this decade, companies confront the need to wring unnecessary buffers of time out of their development processes without destroying creativity. The ability to develop new products and processes quickly and effectively by identifying the intersection of technological potential with user benefit will distinguish industry leaders. Truly understanding user needs—not just their demands—enables speed at no sacrifice of inventiveness.

Notes

1. Such techniques are often used as the front end for quality function deployment processes that translate user requirements into engineering specifications. See John Hauser and Don Clausing (1988).

2. This danger is exacerbated if senior officials try to micromanage technical project details. In one company studied by Leonard-Barton, project teams frequently complained that senior management "swooped down" midway through projects to "tinker" with technical details that they were no longer in a position to judge. Senior managers can also be personally invested in an obsolete technology. In another company, a major move from glass to plastic materials was delayed at huge cost to the corporation. The principal reason for the delay, according to company informants, was that one senior official could not bear to desert his primary skill base and expertise in glass-making. Only when he left the company—some years after competition had already proven the profitability of the move—was the switch finally made.

3. Of course, either of these may be preceded by a process of discovery. However, technological discoveries only provide opportunities; the discoveries must be shaped into useful products and processes.

References

Bacon, Glenn, Sara Beckman, David Mowery, and Edith Wilson. 1994. "Managing Product Definition in High Technology Industries: A Pilot Study." *California Management Review* 36: 32–52.

Birnbaum, Joel. 1992. Interview by Dorothy Leonard-Barton, November.

Booz-Allen and Hamilton, Inc. 1982. *New Product Management for the 1980's.* New York: Booz-Allen & Hamilton, Inc.

Bowen, Kent, Kim Clark, Charles Holloway, and Steven Wheelwright, eds. 1994. *The Perpetual Enterprise Machine: Seven Keys to Corporate Renewal through Successful Product and Process Development.* New York: Oxford University Press.

Bower, Joseph, and Thomas M. Hout. 1988. "Fast-Cycle Capability for Competitive Power." *Harvard Business Review* (November–December): 110–18.

Brown, John Seely, and Susan E. Newman. 1985. "Issues in Cognitive and Social Ergonomics: From Our House to Bauhaus." *Human-Computer Interaction* 1: 352–91.

Burgelman, Robert. 1983. "A Process Model of Internal Corporate Venturing in the Diversified Major Firm." *Administrative Science Quarterly* 28: 223–44.

Calantone, R., and Robert G. Cooper. 1979. "A Discriminant Model for Identifying Scenarios of Industrial New Product Failure." *Journal of the Academy of Marketing Science* 7: 163–83.

Christensen, Clayton. 1992. "The Innovator's Challenge: Understanding the Influence of Market Environment on Processes of Technology Development in the Rigid Disk Drive Industry." Ph.D. diss., Harvard Business School.

Cooper, Robert G. 1975. "Why Industrial New Products Fail." *Industrial Marketing Management* 4: 315–26.

———. 1986. *Winning at New Products.* Reading, Mass.: Addison-Wesley.

Cooper, Robert G., and Elko J. Kleinschmidt. 1986. "An Investigation into the New Product Process: Steps, Deficiencies, and Impact." *Journal of Product Innovation Management* 3: 71–85.

———. 1990. "New Product Success Factors: A Comparison of 'Kills' versus Successes and Failures." *R&D Management* 20, no. 1: 47–63.

Dougherty, Deborah, and Trudy Heller. 1994. "The Illegitimacy of Successful Product Innovation in Established Firms." *Organization Science* (May): 200–18.

Doyle, John. 1985. "Commentary: Managing New Product Development: How Japanese Companies Learn and Unlearn." In *The Uneasy Alliance: Managing the Product-Technology Dilemma,* edited by Kim Clark, Robert Hayes, and Christopher Lorenz. Boston, Mass.: Harvard Business School Press.

Dumaine, Brian. 1989. "How Managers Can Succeed Through Speed." *Fortune* 119 (13 February): 54–74.

Finlay, Douglas. 1991. "Office Printers: Higher Quality, Lower Prices." *Office,* 113, no. 5: 42–44.

Foster, Richard. 1982. "Organize for Technology Transfer." *Business Week* (24 May): 24–33.

Freeze, Karen, and Dorothy Leonard-Barton. 1991a. *GE Plastics: Selecting a Partner.* Boston, Mass.: Design Management Institute.

———. 1991b. *Polymer Solutions: Tempest About a Teapot.* Boston, Mass.: Design Management Institute.

Gupta, Ashok, and David Wilemon. 1990. "Accelerating the Development of Technology-Based New Products." *California Management Review* (Winter): 24–44.

Hauser, John R., and Don Clausing. 1988. "The House of Quality." *Harvard Business Review* (May–June): 63–73.

Hayes, Robert. 1985. "Strategic Planning—Forward in Reverse?" *Harvard Business Review* (November–December): 111–19.

Hitt, Michael, and R. Duane Ireland. 1985. "Corporate Distinctive Competence, Strategy, Industry, and Performance." *Strategic Management Journal* 6: 273–93.

Hof, Robert D. 1993. "Hewlett-Packard Digs Deep for a Digital Future." *Business Week* (18 October): 72–75.

Hopkins, D. S., and E. L. Bailey. 1971. "New Product Pressures." *Conference Board Record* 8: 16–24.

Kanter, Rosabeth Moss. 1988. "When a Thousand Flowers Bloom: Structural, Collective, and Social Conditions for Innovation in Organizations." In *Research in Organizational Behavior,* edited by Barry M. Staw and L. L. Cummings. Vol. 10. Greenwich, Conn.: JAI Press.

Kantrow, Alan. 1986. "Wide-Open Management at Chaparral Steel: An Interview with Gordon E. Forward." *Harvard Business Review* (May–June): 96–102.

Leonard-Barton, Dorothy. 1987. "The Case for Integrative Innovation: An Expert System at Digital." *Sloan Management Review* 29, no. 1: 7–19.

———. 1989. "Implementing New Production Technologies: Exercises in Corporate Learning." In *Managing Complexity in High Technology Industries: Systems and People,* edited by Mary Von Glinow and Susan Mohrman. New York: Oxford University Press.

———. 1992a. "Core Capabilities and Core Rigidities: A Paradox in Managing New Product Development." *Strategic Management Journal* 13: 111–25.

———. 1992b. "The Factory as a Learning Laboratory." *Sloan Management Review* 34, no. 1: 23–36.

———. 1992c. "Inanimate Integrators: A Block of Wood Speaks." *Design Management Journal* 2, no. 3: 61–67.

———. 1995. *Wellsprings of Knowledge: Building and Sustaining the Sources of Innovation.* Boston: Harvard Business School Press.

Leonard-Barton, Dorothy, and Deepak Sinha. 1990. "Dependency, Involvement, and User Satisfaction: The Case of Internal Software Development." Working paper #91-008, Harvard Business School, Cambridge, Mass.

Maidique, Modesto A., and Billie Jo Zirger. 1985. "The New Product Learning Cycle." *Research Policy* 14, (December): 229–313.

Prahalad, C. K., and Gary Hamel. 1990. "The Core Competence of the Corporation." *Harvard Business Review* (May–June): 79–91.

Rothwell, Roy, et al. 1973. Project SAPPHO—A Comparative Study of Success and Failure in Industrial Innovation. London: Centre for the Study of Industrial Innovation.

———. 1974. "SAPPHO Updated—Project SAPPHO Phase II." *Research Policy* 3: 258–91.

Rumelt, Richard. 1986. Reprint. *Strategy, Structure and Economic Performance.* Harvard Business School Classics. Boston: Harvard Business School Press. Original edition, Boston, Harvard Business School Press, 1974.

Schwartz, Peter. 1991. *The Art of the Long View: Planning for the Future in an Uncertain World.* New York: Doubleday Currency.

Serum, James. 1992. Interview by Dorothy Leonard-Barton, November.

Shiba, Shoji, Richard Lynch, IRA Moskowitz, and John Sheridan. 1991. "Step by Step KJ Method." CQM Document No. 2, The Center for Quality Management, Analog Devices, Wilmington, Massachusetts.

Souder, William. 1987. *Managing New Product Innovations.* Lexington, Mass.: Lexington Books.

Stalk, George Jr. 1988. "Time—The Next Source of Competitive Advantage." *Harvard Business Review* (July–August): 41–51.

von Braun, Christoph-Friedrich. 1990. "The Acceleration Trap." *Sloan Management Review* 32, no. 1: 49–58.

von Hippel, Eric. 1983. "Novel Product Concepts from Lead Users: Segmenting Users by Experience." Working paper 1476-83. Sloan School of Management, Massachusetts Institute of Technology.

———. 1988. *The Sources of Innovation.* New York: Oxford University Press.

Wilson, Edith. 1990. Product Definition Factors for Successful Designs. Master's thesis, Stanford University.

9

RETHINKING THE ROLE OF INDUSTRIAL RESEARCH

Mark B. Myers and Richard S. Rosenbloom

M ANY FIRMS HAVE been driven by the dual forces of increasing competition and accelerating technological change to renew themselves through initiatives that introduce new "organizational architectures" and "reengineer" the way in which they do business.[1] These initiatives have forced managers to focus on gaining a better understanding of fundamental business practices in order to adapt the practices to changing needs of the corporation. In technology-based enterprises, the management of innovation is a key business process. For these firms, competitive advantage is derived from a continuing stream of innovations, not only in new products and services but also in the distribution and support of those products and services. Because research constitutes a vital part of the innovation process, senior management needs to rethink the role of research.

This management task is especially important to large, mature, technology-intensive companies—AT&T, Digital Equipment, IBM, Kodak, NEC, Philips, and Xerox, for example—as they face growing global competition. As rapid change in technology alters markets, destabilizes competitive balances, and offers opportunities for competitive repositioning, large enterprises must strengthen their capacity to create a continuing stream of productivity-enhancing innovations in products, services, and work practices. It will require more than downsizing and restructuring to meet the challenge of global competition effectively. These companies must seek to restore the high levels of innovativeness and competitiveness that made them great in the first place. For such firms, then, repositioning the research function as part of the firm's "engine of innovation" is an important priority. This task is complex because the "engine of innovation" must be capable of operating on two planes. Beyond the continuous enhancement of established products and services and constant reduction in their costs, the "engine" must be capable of commercially exploiting new technologies that will give rise to new classes of customers, channels, and uses. Hence the firm will need the capacity to compete effectively for new and emerging markets. In competition focused at the confluence of emergent markets and technologies, the need to cope with the dynamics of their interactions in the face of the consequent uncertainties requires a different set of organizational capabilities.

Because the field of competition in technology-based markets is already global and the location of distinctive capabilities is increasingly dispersed, innovative capacity is going to depend on dynamic linkages among geographically distributed nodes of specific capabilities. Innovative firms will need the ability to form competitive alliances on a global scale to complement their own capabilities through partnerships or licenses. Although they are intangible assets, intellectual property and core competence make up an important currency for the formation of dynamic global alliances. As a firm evolves toward a global network of capabilities, these assets become not only the basis of both radical and ongoing innovation but also the main currency for the formation of effective alliances.

The firms most affected by these new competitive requirements include those that account for the lion's share of research in industry today. Consequently, the change processes just identified present both threats and opportunities to the world's principal industrial research establishments, many of which have already had their sense of value and order threatened by corporate reexaminations of fundamental assumptions about the contributions of research to a firm's bottom line. Furthermore, a new level of financial stringency—budget cuts as well as demands for greater relevance—is being imposed on many organizations that have long measured appreciation by the size of their annual budget increases. On the other hand, rethinking the role of research in order to adapt to a new environment can bring new opportunity to the industrial research community.

In a world of accelerating change, the research organization is uniquely positioned to identify the salient forces of change and to help shape the firm's strategies in anticipation of their consequences. This involves not only a vision of future technological possibilities but also an interpretation, made in conjunction with operating units, of the implications of these possibilities in the context of present and future markets and work processes. Because it operates with a longer time horizon than most corporate units, the research organization brings a distinctive perspective to the interpretive task. The organization also can make distinctive contributions to the initiation of strategic action. Because research operating budgets usually permit research managers more influence over the commitment of strategic resources than most business unit managers have, research can invest in the development of novel technology platforms in anticipation of new needs.

The pursuit of these goals implies acceptance of a changed role for the traditional corporate research organization. In today's environment, research must see its role in the context of the whole corporation, its markets, its customers' needs, and its core competencies. The organization must be able to read and interpret technological and sociological forces of change in the context of the strategic needs and capabilities of the corporation. Furthermore, it must see its success in the context of how well the corporation succeeds within dynamic and highly competitive markets.

In light of this, this chapter will explore the idea of a new self-concept for industrial research. In the first section, we will sketch a "total process" view of innovation in which there is continuing interaction between the dynamics of changing markets and the possibilities inherent in new technology. It is time to move beyond long-standing debates about the value of basic or applied and long-term or short-term research. These are characteristics of the old linear model of innovation. Industrial research should be seen as the structured acquisition of knowledge and support for learning across the total process of creating, making, selling, and supporting innovative and efficient products and services. It should be guided, we argue, by a vision of technology that is well integrated with the larger strategic vision for the firm as a whole and with the current and emerging realities of major marketplaces.

If the corporation is to be effective as an "engine of innovation," the research organization has an integral role to play. The research organization can help to align the corporation's strategic vision with emerging technological opportunities and constraints and should play a leading role in bringing the corporation's organizational capabilities to a position consistent with its strategic vision. The final section then discusses some of the implications of the total process view of innovation for research managers.

INNOVATION AS A TOTAL PROCESS

To define the role of research management in industry today, we must rethink the way in which innovation occurs in a large technology-based corporation. To do this, we must abandon the implicit assumptions of a linear view of innovation. As displayed in Figure 9-1, the linear model of innovation implies that technology flows to the marketplace in much the same way that a relay race is run. In this view, technology is the baton passed among different groups that are separated organizationally, geographically, and chronologically. This view is an abstraction of a process that was probably never representative of the innovation processes of organizations. While there have been many critics of the linear model,[2] its assumptions are still firmly embedded in the conceptual language of both the research organization and the corporation. We speak of technology transfer in a time-linear fashion. We distinguish basic, or long-term research from applied, or short-term, research by its time distance from application. Corporations separate their researchers organizationally and physically from their markets and from the ultimate consumers of technology. While the linear view is out of fashion in discussions of innovation, its assumptions are embedded in many aspects of practice and there is no consensus supporting a new innovation model to replace it.

In our view, the realities of innovation in industry are best conceptualized in the "chain-linked" model proposed by Stephen Kline and Nathan Rosenberg (1986). In this model, illustrated in Figure 9-2, the central "chain of innovation" which starts and ends with markets rather than with events in science or technology, is just one element. Of comparable importance, but more complex and diverse in character, are the multiple links to research and knowledge and the feedback loops that integrate design, test, production, distribution, and service functions within the central chain. In this model, innovation appears as a complex mixture of parallel and serial processes. Kline and Rosenberg thus recognize a reality that has always been recognized by managers of research: Innovation is a complex and often disorderly process with varied and diverse manifestations. In brief, "there are

Figure 9-1. The Linear Model of Innovation

Figure 9-2. Chain-Linked Model

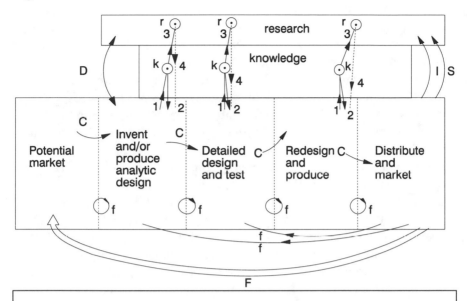

KEY: Symbols on arrows:

C = Central-chain-of-innovation

f = Feedback loops

F = Particularly important feedback

k-r = Links through knowledge to research and returns paths. If problem solved at node k, link 3 to r is not activated. Return from research (link 4) is problematic—therefore, dashed line.

D = Direct link to and from research from problems in invention and design.

I = support of scientific research by instruments, machines, tools, and procedures of technology.

S = Support of research in sciences underlying product area to gain information directly and by monitoring outside work. The information obtained may apply anywhere along the chain.

Source: Slightly modified from Stephen J. Kline and Nathan Rosenberg. "An Overview of Innovation." In *The Positive Sum Strategy*, edited by Ralph Landan and Nathan Rosenberg. Washington, D.C.: National Academy Press, 1986. Copyright 1986 National Academy Press. Reproduced by permission.

many black boxes rather than just one" (Kline and Rosenberg 1986, p. 280).

In this characterization of innovation, the central process is design rather than science or technology. As Kline and Rosenberg explain, "Successful innovation requires a design that balances the requirements of the new product and its manufacturing processes, the market needs, and the need to maintain an organization that can continue to support all these activities effectively" (p. 277). In this model, then, design is carried out in an economic context, and the forces driving innovation are, for the most part, the economic parameters of the marketplace rather than the institutional imperatives of science and technology. Although science and technology in this model rarely are drivers of the process of innovation, they are essential enablers. Therefore, science is depicted in two constituent parts: bodies of "stored knowledge" and processes of research. Each part enters into the innovation process across its breadth. Thus, science and technology interact with innovation in multiple ways, including learning from previous market engagements, continuing enhancement of innovative capabilities through the creation of new tools and instruments, and acquisition of new knowledge from the external technical infrastructure. The innovation process in this view is primarily a learning process in which knowledge has a central role as the key ingredient and the principal output. Thus, as Kline and Rosenberg say, "the central dimension that organizes innovation, if there is one, is uncertainty" and the process of innovation can be viewed "as an exercise in management and reduction of uncertainty" (pp. 274–75).

The chain-linked model, as we understand it, is functional rather than organizational in concept. That is, the boxes in Figure 9–2 denote generic activities, not generic organizational entities. The box labeled "research," then, indicates a kind of work rather than the "Research Organization." As the model implies, research (with a little "r")—meaning the structured acquisition of knowledge—may play a role at any point in the innovation process and may occur within any organizational entity. Clearly, the Research Organization (with a big "R"), as the entity charged with carrying out the firm's principal research tasks, can play a larger role in nurturing and encouraging the use of research skills throughout the organization, especially in relation to innovative activities.

In our view, the Kline-Rosenberg model captures the complex nature of the innovation process in a generic sense and holds some important implications for research managers. However, to explore the dynamics of the research and technology organizations within a firm, it is necessary to extend the model. The extensions we propose involve the explicit recognition of (1) organizational capabilities and (2) the special characteristics of innovations built on discontinuities in technology or markets.

Organizational capabilities are the foundations of competitive advantage in innovation. Three are worth noting:

1. Firm-specific knowledge. This represents the accumulated learning of the organization. It is the set of understandings pertinent to the business and its technologies that goes beyond what would be known by a journeyman scientist or engineer in a field. Thus it is distinguished from the body of generally accessible knowledge. The specific knowledge of a firm is embodied in the firm's people and its technology platforms, products, and processes.

2. Communities of practice. These are ensembles of skilled technical people with common and complementary expertise working across the organization over sustained periods of time at certain defined tasks, such as the design of man-machine interfaces or the control of process variation. These communities span organizational divisions and provide both a repository for the firm's expertise and a medium for communication and application of new knowledge.

3. Technology platforms. These are an output of the design process; they provide a common framework on which families of specific products and services can be created over time. Depending upon the industry, a platform may be expressed in the design of a core product, a critical process, or the organizational routines of a service operation. A platform comprises an ensemble of technologies configured in a system or subsystem that creates opportunities for a variety of outputs. Hence a polymerization unit in a chemical plant may support a range of specialized products. The architecture of a platform, then, is an important element; opportunities are multiplied if the platform permits ready reconfiguration and the easy substitution of important elements. Sony's Walkman product line is based on a few design platforms that support hundreds of product variants (Sanderson and Uzumeri 1990).

Figure 9-3 is a graphic representation of the chain-linked model expanded to represent these elements.

What happens in this model when radically new technology created by a research organization is the basis of innovation? Radical innovations, the first synthetic polymers for textile use or the laser printer, for example, provide the need and opportunity for researchers to contribute to the design of a novel platform. They also create situations of high uncertainty for the management of innovation. As Kline and Rosenberg point out, "Like fundamental research, radical innovation is inherently a learning process" (p. 297). Because emergent technologies often lead to new or emergent markets, the management of inno-

Figure 9-3. A "Total Process" Model of Innovation

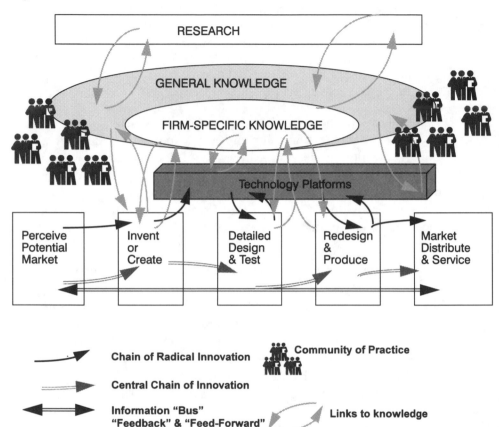

vation must deal with the uncertainty inherent in both the technology and the new or emergent market as well as with their interactions.[3] For a firm to seize a competitive advantage in high technology fields, however, it must have the ability to explore and exploit the coupled dynamic of emergent markets and emergent technologies. Thus, the renewal of a firm's innovative capabilities implies a strengthening of the role of research in the central chain of innovation (in the Kline-Rosenberg sense) that leads to new technology platforms. In tracing the implications of this, the limitations of the linear model of research's role in innovation and the need for a total process view as expressed in the chain-linked model become clear. Unfortunately, Kline and Rosenberg limit their discussion to technological uncertainties. Research management must deal with both technological and market uncertainties and with their interactions. In the following section we will suggest an approach of "radical incrementalism" for handling these uncertainties and their interactions. Under this approach new technical concepts are continually tested within a larger vision of future markets through a series of practical learning experiences of limited scale with customers. Such an approach requires the creation of technology platforms with architectures that enable rapid reconfiguration and extension across markets.

THE ROLE OF RESEARCH

Competing through innovation as a total process emphasizes the effectiveness of the whole organization. Although new technology advances are essential, they alone are not sufficient. Exciting discoveries may become scientific artifacts if the organization cannot relate them to customer needs and then design, manufacture, distribute, and support the new products and services they enable. The experience of Xerox Corporation illustrates the shortcomings of technological leadership when it is not matched by overall innovative performance. Although its research organization made many important contributions to the base reprographic business throughout the 1970s and 1980s, in the early 1980s, the corporation gained a reputation for "fumbling the future" as research created a series of innovative concepts that yielded only modest economic benefit to the company but were highly successful in the marketplace when commercialized by others (Smith and Alexander 1988). In the late 1980s, William J. Spencer, the head of research, directed an analysis of three significant products based on Xerox technology leadership. The analysis hoped to determine processes in the management of technology that could be used to avoid the repetition of past failures. Xerox researchers were among the first to conceive and develop the technologies for the desktop laser printer (later pioneered by Canon and Hewlett-Packard); the related printer page description language (exploited with huge success by Adobe); and the desktop workstation (for which Apple, IBM, and Sun, among others, found huge markets).

The analysis identified several factors contributing to the company's failure to exploit these opportunities successfully. A heavy emphasis on technology push rather than market pull contributed to the failure to couple marketing and technology effectively in understanding and addressing the customer's real work processes and requirements. In the case of the workstation, Xerox had assumed a system configuration that required a large initial investment for the installation of a fully networked client-server implementation well before the general user understood or sensed the need for it. The closed architecture of the system limited extendability and creative contributions by third-party vendors. Although the architectural concepts have proven correct, they were not connected to the perceived customer needs of the time. The market was developed by others who served the customer with simpler or focused implementations of the core concepts, such as the less-expensive PC platforms that supported such market-forming applications as spreadsheets. Adding to the problem of technology push rather than market pull was the unevenness of the company's capabilities across the varied elements of the innovation process—technology, design, manufacture, and sales and service for a new or emergent customer base. Perhaps most important, overall, was a lack of clarity about the role of research in the innovation process and the failure to integrate research effectively into the processes by which the corporation made strategic commitments.

What are the capabilities that will enable a corporation to use its research organization in a broader role in the innovation process? In the most basic terms, the corporation needs a technology strategy that aligns the "where, when, what, and how" of innovation. When innovation is a total process (as sketched in Figure 9-3), the direction and velocity of the firm's innovative efforts will be defined, respectively, by its strategic vision and its organizational routines and work practices. By defining where the firm is heading, a strategic vision shared among research and operating leadership identifies where the firm will need innovative science and technology. By establishing the capacity for high-velocity innovation across the organization, new work practices can provide the rapid time to market that is necessary to exploit the fruits of research profitably in today's competitive environment.

Given these foundations, how does the research organization contribute most effectively to the profitable execution of the chains of innovation mapped in Figure 9-3? *Effective* research will contribute to the base of general knowledge, but *productive* research requires the corporation to build the organizational capabilities—firm-specific knowledge, communities of practice, and technology platforms—needed to realize its strategic vision. While a full discussion of the dilemmas inherent in formulating a research strategy in today's environment is beyond the scope of this paper, we will comment briefly on the "where, when, what, and how" of technological innovation and the use of "radical incrementalism" as a philosophy for implementing ideas effectively when radically new technologies intersect with emerging markets.

Vision

Technology strategy should begin with a powerful research vision that is integrated into the corporate strategic intent. A shared vision gives unity and purpose within the research community and helps to establish the technical foundation of the business strategy. The vision anticipates the interaction of technology trends with economic, competitive, and social trends in the same time frame. The vision is large in scope, yet it is definable and actionable. It has an evolving time horizon in which details of the landscape come from probing and learning. Finally what comes from this vision and the research that supports it are major new insights and ideas that impact the future products, services, and internal work practices of the organization.

It is the responsibility of the research and technology organization to read the technology forces of change; to interpret them in the context of present and future markets and corporate work processes; and to stimulate strategic actions in response to them. To illustrate the challenge of this task, consider the recent acceleration in the rate of change in the speed of computation and communication technologies, as shown schematically in Figure 9-4. As their speed increases, the technologies of computation and communications are converging. This development is leading to new business paradigms because transactions can take place around the world at any time, between any places, without physical shipments, and because mass-customization creates economic market segments of one (Davis 1987). All corporate research organizations should be envisioning and exploring the impacts and opportunities that converging technologies will create in their present and future marketplaces.

The Time Dimension

Because the total process model of innovation both starts and ends with markets, two important issues arise with respect to time-based management. The first deals with the inherent uncertainty of the timing of market developments and the second with the ability to respond rapidly to emerging market opportunities. The inherent unpredictability of the details of markets and the requirements of those

Figure 9-4.   Accelerating Pace of Change of Computation and Communication

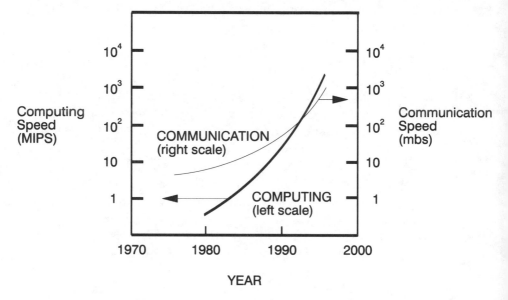

Note: The left axis measures the increase in personal computer power at the individual workstation in millions of instructions per second ["mips"]; the right axis measures the increase in bandwidth in local area networks in millions of bits per seconds ["mbs"]. This is derived from many "white board conversations" with John Seely Brown of Xerox Palo Alto Research Center in 1990.

markets presents a challenge to those managing the timing of commercialization of significant innovations. Economic, social, political, and technological forces can change emerging markets. Who would have predicted the details of the political landscape of Europe or the present characteristics of the world economy two, four, or six years ago? The details of markets are dynamic, and therefore, the core competence for exploration of emergent markets and emergent technologies must also be dynamic.

While the direction of technology change can be anticipated, this is not sufficient for success. The dynamic skill that is required is the ability to be positioned well when technology change leads to the emergence of a market. It is the ability to read the time confluence of economic, social, political, and technological forces, which are like small tributaries coming together into a swelling stream, that is vital to the corporation. Some technology initiatives fail; others prosper and sometimes converge. In retrospect, one may see long gestation periods for successes, but their potential is not always clear in real time, as Figure 9-5 illustrates. At time *a* one can observe the emergence of several independent technological developments; by time *b* some have already diminished, others are growing in importance, new ones are emerging, and the beginnings of a market can be seen; by time *c* the technologies are linked, and market activity is in a strong growth phase.

The FAX revolution of the early 1990s is one example of a confluence of trends. The underlying technologies to support the communication of images by facsimile were available in the late 1960s. Xerox had tried to develop a market at that time but was unable to create a profitable business. It was the right idea, but it was promoted too early in the stream of emerging technologies. The next twenty years brought the development of a number of technology advances that together created conditions for a broad-based market. These advances included: compression, enabling higher speed; plain paper marking technology, enabling enhanced image quality; incessant cost-performance improvement of silicon electronics technology, enabling automation, integration and a more than 90 percent cost reduction; standards, enabling universality; and improved user interfaces, enabling ease of use. The relative attractiveness of facsimiles was further en-

Figure 9-5. The Confluence of Trends

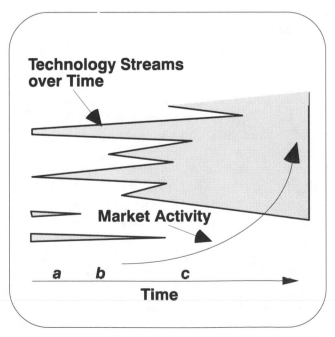

hanced by dramatic increases in the costs of surface mail, which were accompanied by decreases in its reliability and reductions in long-distance telephone tariffs due to competition and technology. In the timing of innovation, there is the risk of being too early as well as too late.

Once the confluence of events is right for a market to emerge, the ability to respond rapidly is essential. This is the second time management issue. The time between the identification of a potential market and the creation of the ability to serve it must be as short as possible. Fast clock speed in this process is a stepping-stone to competitive advantage. Longer cycle functions in research, such as developing innovative technology platforms and the capabilities of communities of practice, must function as parallel streams to short-cycle activities. As a popular discussion of time-based management points out, fast innovators can experiment with their customers as they fine-tune innovation. Experience has shown that companies that have significantly reduced their innovation time have distinctly different patterns of innovating and competing than companies that have not (Stalk and Hout 1990).

Technology Platforms

What kinds of requirements does the time dimension place on the development of technology? Clearly, it is necessary to have reconfigurable platforms, or ensembles, of technologies that can be used to build prototype systems, which then engage the customer in coproduction. We can envision these platforms as building block constructions. The technologies are modular blocks that can be put together in a systematic but open way. That is, they have an extendable architecture. The architecture must be so extendable that it permits reconfiguration for specific applications as well as the upgrading of individual modules over time.

The platforms are derived from sustained technology investments that are guided by the research vision and bounded by the strategic intent of the business. Their creation should represent a dynamic and creative interplay of the possibilities of emergent technologies and emergent markets. This interplay leads to a set of technology bets on future customer requirements and the identification of the critical questions that give both the researcher and the marketer directions for further investigation. The process must be interactive and dynamic in time. The research vision must be continuously tested by an experiential reading of the technological and sociological forces of change.

The scope of the research investment will be larger than the eventual technology ensembles employed in the product platforms because it must serve three important needs. The first need is the need to map the full extent of the technological frontiers that are important to the realization of the corporate business vision and intent. Thus, these investigations will encompass the range of technologies to which the corporation will make value-added commitments—invention, design, manufacturing, and service—and involve technologies in which the corporation will want to have knowledgeable access through purchase or partnerships. The second need is the need to provide qualified technology alternatives for the selection phases of the development funnel described by Wheelwright and Clark (1992). The third need is the need to create a portfolio of marketable intellectual property, which will provide currency for the establishment of strategic partnerships to assure market access.

The research program must be supported by the capabilities required to cover the entire extent of the innovation process. Thus, the most productive research organizations will be those communities that have a multitude of interests as well as a variety of disciplines. That is, staffs of these communities will include persons possessing traditional technical skills in sciences and mathematics and complementary interests in such fields as economics, languages, design, manufacturing processes, and so forth. The best organizations will be those that can attract and hold people who are challenged by the prospect of moving through many fields of knowledge during their careers.

Concurrent with the need to develop reconfigurable technology platforms is the need to identify and develop across the entire organization the broad range of specific knowledge and communities of practice that are required to develop, produce, sell, and service products based on advances in technology. The identification and engagement with the operating units to coproduce these core skills of the organization may be one of the most important contributions that the research function can make to the innovative capabilities of the firm.

RADICAL INCREMENTALISM

Given the inherent unpredictability of the details and timing of future markets, corporations need a practical approach to exploring the dynamic of emergent markets and emergent technologies. *Radical incrementalism,* as we call it, is an approach that tests radically new concepts in a series of practical market learning experiences (see Figure 9-6). The approach goes beyond *kaizen,* or gradual and continuous improvement within existing markets. Through a succession of "bite size" experiences, radical new ideas and new market needs are brought together. The learning that occurs in each experience is taken back to the laboratories and used to modify and improve the product for the next application event.

A powerful variant of this approach involves the codevelopment teams that were discussed in chapter 4. As noted there, a codevelopment team works with a *lead* customer to come to a shared understanding of an important problem that may be solved by an emergent technology. Based on the team's understanding of the customer's requirements, a working solution is designed and implemented. This system is evaluated through use, which usually elicits underlying requirements that were not articulated in the first round. This leads to a new round of redesign and evaluation. Thus, together the team and the customer develop a solution to the problem. Lead customers are chosen on the basis of the opportunity they present to refine and flesh out the vision guiding the innovative effort. By the nature of their problem and the quality of their enterprise, working with them can present a major learning opportunity for advancing the state of the art of a business solution.

Ultimately codevelopment teams offer a way to explore emergent markets and emergent technologies. This helps to produce products

Figure 9-6. Radical Incrementalism

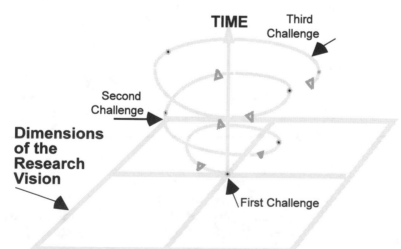

that are naturally attuned to real world product environment integration and thus serves to "pull" rather than "push" new technologies.

CONCLUSION

Obviously, the best route to adaptation of the role of research depends upon the particular circumstances, history, traditions, industry, and national setting of the firm. A North American specialty chemicals producer, a European pharmaceutical house, or a Japanese computer maker will see different degrees of urgency and different tasks to be accomplished. Without implying any diminution of the importance of the particulars of each company's situation, we shall conclude with a few brief comments on some possible common themes.

First, in today's global competitive arena, renewal and adaptation of the research establishment is an important goal for any firm investing significantly in scientific research. If the firm is sufficiently large, affluent, and dependent on technology to have a research organization, and if the firm is in a field in which technological opportunities are strong enough to promise results from such an organization, the firm is sure to be in an environment in which continuing adaptation to change is required. In other words, almost by definition, research managers are within firms that must exhibit the capabilities of a learning organization.

Second, as the intellectual capital underlying the firm's core capabilities becomes increasingly important competitively and as the need for continuous improvement in that capital stock becomes more essential, it becomes increasingly critical for research to become an organic part of the mainstream of the value chain in the business. Putting this in a different way, in today's world, the old notion of research set apart from operations and customers is obsolete. Research can strengthen the value chain through three vehicles: a technology *vision*—vision that fuses with the firm's marketplace vision to guide strategic direction, as well as detailed intelligence to guide strategic commitments on an ongoing basis; *platforms and tools* for change, not only new technologies at the core of innovative products but also new methods to raise the productivity of knowledge workers throughout the value chain; and *people* who flow from research into key organizational roles, providing channels back to research for learning from operations and customers, and bringing to other units the capacities for learning new skills that are inherent in the research function.

Just as research activities must become more integrated into the value chain, research managers must become more integral to the direction of the business. The corporate strategic vision must synthesize visions of future realities in both markets and technologies. Although the 1970s vision of information technology at Xerox PARC was powerful and right, the gap between PARC and the core commitment process of the corporation blocked the company from reaping abundant opportunity. To get the right synthesis, at least two things are needed. First, the CEO must provide support and require it from the firm as a whole. Second, the research leadership must deepen its understanding of customer needs, manufacturing realities, and other business requirements.

As research becomes increasingly embedded in the value chain, it seems likely that that leading industrial research enterprises will compete as the nuclei of global strategic networks of firms. Even the largest research organization can no longer expect to own all of the capabilities required for competitive advantage worldwide in both mature and emerging markets. The requisite assets will have to be assembled through a series of shifting alliances with partners large and small. A critical asset for leadership in such networks will be the intellectual capital necessary to define what is needed and to manage effectively the complex relations among allied firms.

Third, our analysis implies that current circumstances call for a step forward in research productivity of a certain kind. It is not a matter of striving for more or better discoveries and inventions but of learning how to produce these in a manner that is useful to those who must make and sell the resultant products and services. The direct involvement of research staff with operating personnel will help to diffuse knowledge of new technologies rapidly and effectively into the operating units.

Finally, corporations restructure when they need to do things in different ways. Changing the form and scale of an organization, however, is inadequate if the people lack the capabilities required for different tasks. One characteristic of a healthy research organization is that its people invest continually in self-renewal. Research can help to establish processes and frameworks for the continuous self-renewal of knowledge workers from engineering to sales and service.

In summary, the competitive paradigm of the future will be based on an organization's ability to identify and participate in changing and emergent markets. The change in markets will be fueled by the continuing increase in the power of technology and reduction in its cost. The challenge arises from the accelerating rate of technical change, as well as from the magnitude of change. An organization's ability to perform in this environment will determine it's future prosperity. The research organization is charged with management of the "future time" of the corporation, and part of that charge is to condition the organization to take advantage of change.

Notes

1. On this topic, see, for example, Michael Hammer and James Champy (1993) and Robert Howard (1992).

2. See, for example, William J. Price and Lawrence W. Bass (1969).

3. The multiple and complex issues posed by radical innovations that may displace core technologies are pertinent here, but a full discussion of them would require a separate paper. Contemporary business events testify to the problems that ensue when a highly successful proprietary platform is challenged by a radically different alternative. One might note such cases as IBM's mainframe architecture and Digital Equipment's VAX in the face of networked RISC architected workstations; Kodak's silver halide imaging against electronic [CCD] imaging for camcorders and still photography; and Xerox's light lens copier platforms challenged by digital reprographics.

References

Davis, Stanley M. 1987. *Future Perfect.* Reading, Mass.: Addison-Wesley.

Hammer, Michael, and James Champy. 1993. *Reengineeering the Corporation: A Manifesto for Business Revolution.* New York: HarperCollins.

Howard, Robert. 1992. "The CEO as Organizational Architect: An Interview with Xerox's Paul Allaire." *Harvard Business Review* (September–October): 107–21.

Kline, Stephen J., and Nathan Rosenberg. 1986. "An Overview of Innovation." In *The Positive Sum Strategy,* edited by Ralph Landau and Nathan Rosenberg. Washington, D.C.: National Academy Press.

Price, William J., and Lawrence W. Bass. 1969. "Scientific Research and the Innovative Process." *Science* 164 (16 May): 802–06.

Sanderson, Susan Walsh, and Vic Uzumeri. 1990. "Strategies for New Product Development and Renewal: Design-Based Incrementalism." Working paper, Center for Science and Technology Policy, School of Management, Rensselaer Polytechnic Institute.

Smith, Douglas K., and Robert C. Alexander. 1988. *Fumbling The Future—How Xerox Invented, Then Ignored, the First Personal Computer.* New York: William Morrow.

Stalk, George, Jr., and Thomas M. Hout. 1990. *Competing Against Time.* New York: The Free Press.

Wheelwright, Steven C., and Kim B. Clark. 1992. *Revolutionizing Product Development.* New York: The Free Press.

CONCLUSION: SHAPING A NEW ERA

Richard R. Nelson, Richard S. Rosenbloom,
and William J. Spencer

THE SIGNS SEEM CLEAR: We are witnessing the end of an era for corporate scientific research laboratories in the United States. The chapters in part 2 illustrate the profound changes that are taking place within these organizations. Research directors have had to direct their research away from fundamental science and pioneering technology toward activities that are more relevant to ongoing product and process development, more likely to produce results that can be kept proprietary, and more certain to produce a commercial payoff in the near future. At the same time, to increase productivity and shorten the time required to bring new technologies to market, they have restructured and resized their organizations. Indeed, some companies have abandoned research entirely.

The forces behind these changes are multiple and powerful. Strong competitive pressures in an increasingly global marketplace have forced greater corporate attention to the bottom line, hence to discretionary budgets like research. Companies once able to dominate their markets, AT&T, IBM, Kodak, and Xerox, for example, now find it difficult to derive a full measure of economic benefit from the technologies emerging from their laboratories.[1] Added to these economic forces are changes in the sociopolitical context in which the laboratories operate. What might be called the "American system" of industrial innovation was built on the strong interplay of scientific institutions in industry, universities, and the public sector.[2] However, the end of the cold war is reshaping the allocation of federal resources for science and advanced technology within that system with undoubted consequences in the laboratories in the private sector.

If nothing else happens, the likely result of driving long-run, pioneering research from companies that have supported it strongly in the past, will be industries in which somewhat leaner American companies compete more effectively in the near term with their foreign rivals.[3]

Foreign companies may or may not partially compensate by increasing their own central research efforts. As they are under much the same kinds of pressures as American firms, it seems extraordinarily unlikely that increases in corporate central research abroad will compensate for the decline of research by American companies. This implies that industry as a whole will fund fewer activities that open up and explore the next generation of technology than it has in recent decades. While leaner American firms may be better able to compete with foreign rivals, industrial technical advance will proceed with a diminished degree of stimulus from corporate research laboratories.

For the last fifty years, the stimulus has been significant. A web of scientific institutions in industry, government, and academia has created the knowledge base underpinning the economical mass production of computer chips, genetically engineered drugs, lasers, and a host of other staples of modern life. These innovations have been a driving force in the growth of companies and in the overall prosperity of the American economy. In the major science-based industries, corporate research has played a distinctive, even unique, role. Like university research, corporate research has focused on fundamental problems, and when it has succeeded, has opened up broad new areas for technological development. Moreover, corporate research was more consistently targeted on problems and opportunities perceived by industry than was university research. And corporations, more than universities, were able to recruit and retain large interdisciplinary research groups.

Today, deterred by the costs and risks of pioneering technology and by the threat of free-riders, corporations are turning away from "big bets" on novel capabilities, but without such investments, where will these corporations find the sources of global competitive advantage in the years ahead? There is a logical fallacy in every firm planning to be a "fast second" with the next new technology.

TECHNICAL LEADERSHIP IN TODAY'S WORLD

If the future prosperity of a firm is dependent on acquiring new technology as the basis for future products and services, the firm's CEO had better recognize the source of that technology.[4] As suggested in part 1, strategic corporate alliances and industry-university collaborations can make important contributions, but their ultimate effectiveness depends on the scientific and technical capabilities within the firm.

Certainly, to the extent that the firm's own laboratories are expected to continue as generators of novelty, the CEO should be sure that otherwise justifiable resizing and redirection do not impair or eliminate the capabilities required for the identification and development of pioneering technologies.

Perhaps the greatest influence of the CEO lies in his or her responsibility for the choice of technical leadership. "Visionary" technical leadership for corporations may have gone out of style, but CEOs may want to consider the importance of that quality when choosing their candidates for chief technical officer.[5] Mervin Kelly, head of Bell Telephone Laboratories during the 1940s, foresaw critical telecommunications needs and successfully challenged his scientists and engineers to create new technologies like the quartz crystal and the transistor to meet those needs. In preparation for World War II, Kelly persuaded the War Department that superior communication capabilities could be provided by single side-band transmission, a technology then in its infancy. The critical components for this nascent technology were quartz crystal oscillators and narrow-band filters. He then rallied research scientists and engineers to develop the capabilities required to deliver massive numbers of these components. The success of Bell Laboratories in meeting this challenge and the challenge of radar encouraged Kelly to back his scientists in the search for an alternative to the vacuum tube. Promoting the belief that telephony's growth would be stunted without such a device, he challenged his scientists to find a solid-state device capable of low-power amplification. The result was the transistor, yet another example of the fruits of directed research in an industrial setting.

The current circumstances of industrial research seem to call for a different type of technical leadership. Thus, the chief technical officer's job probably will place less emphasis on envisioning and stimulating breakthrough innovations through in-house research and more emphasis on identifying, interpreting, and absorbing technical developments from a wide variety of external sources in industry, government, and academia. Research directors may see some aspects of the new era as instances of "paradise lost." Inevitably, they will have to manage in an environment of increasing resource scarcity even though requirements continue to grow. The research organization will be expected to carry the load of intensifying relationships with university and national laboratories and with corporate strategic allies, while sustaining its own innovative output even as the cost of that output grows more rapidly than the funds budgeted. Part of the answer to these challenges must lie in finding ways to go beyond pressures for resizing and redirection to develop strategies for the renewal of the research organization, as Xerox, IBM, and Alcoa have tried to do (see chapters 4, 5, and 6). In most parts of industry, corporate laboratories are now mature, even "aging," institutions. As Intel and Hewlett-Packard have discovered (see chapters 7 and 8), new thinking can lead to the discovery of new pathways to higher productivity and improved linkages between science and commercial need.

A NATIONAL DILEMMA

The changes occurring within industry as a whole have powerful implications for the national economy. While the pharmaceuticals and biotechnology industries have relied on university research as a primary source of new technology, in a very real sense Bell's Murray Hill Laboratories, IBM's Yorktown Lab, Xerox's PARC, and a few other central corporate research laboratories have played the role of national labs in electronics technology. DuPont's central lab and the laboratories of the other great chemical companies have played similar roles in the evolution of chemical product technology. In both of these industries, particularly in electronics, corporate research has provided the basis for next generation product breakthroughs. Thus, the problems we have described pose a real dilemma for national economic policy and for international cooperation. If we learn to do without these facilities, we risk the loss—or, at least, the weakening—of a force that has been a dynamic factor in the U.S. economy. We can try to resuscitate them, but this will require new ways of funding them, or we can try to replace them with other institutions that can contribute to the economy in much the same way.

Clearly, a national strategy to deal with this dilemma must address both the financial and institutional dimensions of the problem. In our view, the institutional issues are the most challenging. Approaches to funding strategies are largely dependent upon the institutional solutions being pursued. Any new organizational arrangement that evolves must preserve the two features of the great corporate labs that lay behind their effectiveness. First, while focused on long-run and often fundamental research, the work of the labs was strongly informed and molded by perceptions of the needs of industry. Second, the work was large scale and systematic. Workers were able to avoid the distractions inherent in the university setting and to pursue programs with longer time horizons than the typical Ph.D. dissertation project. The best of the industrial laboratories achieved a high degree of coherence of purpose and organizational cohesion, which enhanced their productivity. They could also operate on a scale sufficient to employ experts across a broad range of relevant topics, which facilitates work on many fronts.

It may be possible to preserve these institutional strengths by resuscitating the leading industrial laboratories, but funding is a key issue as the leaders of these corporations will continue to be under strong pressure to reallocate resources to activities that are more likely to contribute to shareholder value (Jensen 1993). Perhaps a solution to the funding problem lies in collective industry funding (possibly with some government help) for these laboratories. Collective industry support of basic research relevant to the industry was the idea behind the Bell Telephone Laboratories prior to the break-up of the Bell system. This type of funding, however, can only work if ownership of the laboratories is shifted from individual companies to industry associations or consortia. Short of that, it is unrealistic to expect firms to be willing to fund research in competitors' laboratories. Another possibility may be the use of government funds, perhaps mixed with industry funds, to keep the extant central corporate research laboratories going, but a significant increase in funding by the federal government would encounter substantial political opposition and, moreover, probably would require that the public funding be "shared around" the firms in the industry. We don't think this is a recipe for good science.

Collective industry support seems more practical for work performed outside the corporation. In both the semiconductor and chemical products industries, a significant number of American companies already tax themselves to provide a pool of money that is used to fund research at universities. For the information industries, institutions like the Semiconductor Research Corporation and Stanford's Center for Integrated Systems are supported by funding from groups of corporate sponsors. Important university programs in biotechnology research are funded by grants or contracts from individual firms. In many of these cases, government funds are also available. Several of the National Science Foundation's engineering research centers have been relatively successful in attracting industry contributions, but the primary source of funding, up until now, has been the federal government.

While the Bell model of a central corporate research laboratory that is owned by and serves substantially all the firms in an industry is attractive, it may be deceptively so. Under the Bell system, a single corporate parent—AT&T—governed both the laboratories and the operating companies. AT&T was able to tax its operating companies to support the Bell Labs, which retained considerable autonomy. It is not clear how many firms would have contributed as much as they did had contributions been voluntary. Since the breakup, Bell Communications Research [Bellcore] has acted as the research arm of those companies, but now the Regional Bell Operating Companies are themselves independent with shared representation on the board that governs Bellcore.

The experience at Bellcore and in laboratories set up by other industrial consortia, for example, the Microelectronics and Computer Technology Corporation (MCC) and SEMATECH suggests that firms that govern the budget allocation process and can decide independently whether or not to contribute, tend to insist on projects with relatively specific and short-run payoffs, if they are successful. This does not bode well for a collective industry-owned basic research laboratory, even if there were matching government funds. Furthermore, Bellcore has encountered other difficulties. In 1995, the Bell operating companies announced their intention to divest their ownership of Bellcore.[6]

An entirely different possibility would be to use government laboratories for industrial basic research. This is not really a radical idea, since the nation has long done this, sometimes quite successfully, in certain fields. The National Institutes of Health have been an extremely successful central corporate research laboratory for the collection of industries concerned with medicine and medical care. A significant fraction of long-run research bearing on agriculture is conducted in government-owned agricultural experimentation stations. The old National Bureau of Standards, now part of the National Institutes of Standards and Technology, has long done basic research of great relevance to industry and worked closely with industry in determining its programs and disseminating its results. Even so, there are two reasons to be cautious about expanding the use of government laboratories for industrial basic research. First, the political and bureaucratic constraints bearing on the operations of government laboratories would make it extremely difficult to get the close interaction with industry that is necessary for this kind of activity to be fruitful. While it would not be impossible, as the preceding examples show, it would be very difficult. Both the National Institutes of Health and the National Bureau of Standards, as well as the agricultural experimentation stations, have been put under considerable political pressure from time to time.

Second, the present is surely a bad time to try to move along this road. With the end of the cold war, many of the weapons-oriented laboratories have lost their missions and soon will be under considerable financial pressure. In the view of those that know the labs well, while a few of them could be converted to supporting civilian industry, most could not. Unfortunately, it is those that could not be converted that have strong political support. It seems very likely that there would be strong political pressures to nominate these labs as substitutes for industrial central research.

Of the available institutional options, then, expanding industrially relevant work in universities seems the most attractive (see chapter 2). However, if universities are to be effective in carrying on the kinds of activities that Bell Labs, IBM Yorktown, Hughes Aircraft, and Xerox PARC are redirecting or abandoning, industry must play a large role in determining the research portfolio and university-industry links need to be made quite strong. Furthermore, the laboratories will need to be sizable, and laboratory management will need to have the capability to focus and coordinate large-scale research efforts. As Rosenberg and Nelson have explained in chapter 2, this means that a university operation would need to stand somewhat separate from the standard university research activities. A significant fraction of the staff would have to be scientists and engineers who are not mainline university faculty members. Still, in the past, universities have had considerable experience operating such laboratories for the service of industry.

While some may question the value of the academic connection if such laboratories have to be somewhat distanced from the heart of the academic enterprise, the connection does offer several important advantages. First, while facilities wholly owned by industry consortia are vulnerable to pressures to shorten time horizons and to concentrate on projects with relatively predictable and specifiable results, in organizations connected with universities these pressures can be offset. Second, over the long run, public support of laboratories affiliated with universities is probably more sustainable than public support of laboratories owned by industry consortia. These two advantages reinforce each other. The fact that funding is from government as well as from industry gives laboratory management a reason and a vehicle for resisting industry pressures to focus on the short run.

Clearly, there are opportunities for developing different kinds of hybrid facilities. For example, industry consortia might designate and largely support one or a small number of university-affiliated laboratories. On the other hand, a university laboratory or a small collection of them could set up long-term relationships with particular industry associations or consortia. It is evident that something like this was in the minds of those who pushed forward the program of university-based engineering research centers. A critical element in the success of these hybrid facilities, however, will be the exchange of technical personnel—students graduating from university and going to industry, post-doctoral fellows working in industrial laboratories, visiting scientists and engineers from industry becoming real participants in university laboratories, and faculty members spending sabbaticals in industrial facilities. Expanding the channels for the exchange of personnel will enhance the interchange of ideas and know-how, which will help industry in the development of new products and processes and universities in understanding the direction of advance most relevant to industry.

SEMATECH has been a pioneer in the exchange of industrial technical personnel; in the mid-1990s there were about 250 assignees from member semiconductor firms working in its Austin, Texas, facilities or in the laboratories of equipment suppliers or in national laboratories. A similar program could successfully transplant industrial scientists and engineers to university laboratories for periods of up to two or three years. In fact, William Spencer has suggested just such an arrangement to deal with the collapse of corporate central research on semiconductor technology. Under his proposal, semiconductor research centers would be established at approximately ten universities. Each center would be dedicated to a specific research "thrust" as defined in the road map sponsored by the Semiconductor Industry Association (see chapter 7). The centers would be staffed by graduate students, faculty, research associates, post-doctoral fellows, and visiting scientists and engineers from the industry. Locating industrial scientists and engineers at university research centers would alleviate some of the problems associated with technology transfer (see chapter 8; Spencer 1990). Spencer characterizes this scheme as the creation of a "virtual Bell Labs" from a network of locations interconnected by modern communications technology, such as the Internet, and coordinated by the framework of the road map.

The National Science Foundation and the Semiconductor Research Corporation have requested proposals for new research centers related to the technologies identified in the semiconductor road map. The initial focus of these centers would be on lithography, interconnect, and environment, safety, and health. The centers would be funded jointly by the semiconductor industry and the National Science Foundation. An issue still to be faced in organizing these centers is whether or not to involve foreign funding. By their very nature, universities are open organizations; this needs to be protected in the future. Since information is readily available on a worldwide basis, there is some rationale for worldwide funding.

There are precedents for university-based centers that provide important national technical capabilities. In the 1950s and early 1960s, materials research centers were funded by the Defense Advanced Research Projects Agency. In the 1960s and 1970s, that same agency funded computer research centers. Recently, as noted earlier, the National Science Foundation has established engineering research centers and science centers. Experience and lessons from these earlier initiatives can guide the creation and operation of centers to serve American industry in the twenty-first century.

LOOKING AHEAD

There are a number of issues that the United States must address in the near future if it is to retain the ability to generate industrial growth based on technological breakthroughs. In our view, the critical issues are institutional rather than financial. They are matters of leadership, collaboration, and priorities. We do not question the adequacy of the resources devoted overall by government and the private sector to technical advance (National Academy of Sciences 1993). However, within the portfolio of efforts funded by the $150 billion currently spent on R&D, a new balance is being struck, possibly at the expense of the nation's ability to build foundations for future technical advance. The issue, then, is one of priorities for spending those funds. This is particularly true of the $75 billion in private funding for R&D. Because the forces producing the current changes are not likely to be easily reversed, new strategies must be sought to retain an appropriate balance between the development of specific products and processes and the creation of new technology.

The nation needs to develop infrastructures and work processes that will be effective in the development of precompetitive technologies (Spencer and Grindley 1993; Smith 1995). Cooperative efforts between firms and among institutions within industry, universities, and government may help in this regard. Among the most promising of these efforts are those joining industrial and university laboratories. It is highly likely that more effective pathways to the exploitation of new technology will encourage greater investment in pioneering efforts. More generally, however, there is a need to strengthen the mechanisms available to coordinate university, government, and industrial R&D

activities. There are a range of possibilities for this sort of coordination, and no single approach is likely to prove effective in meeting the differing circumstances of various fields of industry and technology. What is right for software may not be right for energy. The national government could help to motivate, finance, and coordinate experiments with differing approaches within important industrial sectors.

Because the time of automatic annual increases in R&D budgets is over, it is absolutely vital that decision makers understand technological needs and opportunities. These needs and opportunities must be communicated clearly to industry and government leaders, a task that the scientific and engineering community has not done well in the past. In many fields, the explicit cooperative development of technology road maps should help to educate government and the public about future technology needs and the probable benefits of continued investment in long-term R&D. Industry consortia working with universities and government are in the best position to develop these road maps.

Universities are already under pressure to develop new relationships with industry and to play new roles in the process of technical advance. We encourage them to be responsive to new opportunities for collaboration with public and private institutions. Innovative approaches to the development of intellectual capital will not conflict with their role in developing human capital. On the contrary, these approaches may strengthen the effectiveness of training for technological leadership. It goes without saying that universities must continue to turn out the very best engineers, scientists, and technologically competent leaders possible to manage future U.S. industries.

Finally, within industry, priorities and interorganizational relationships will be particularly sensitive to the quality of leadership. The soundest choices will be made by CEOs who are themselves technically astute or have a visionary trusted advisor as a director of research or chief technical officer. Difficult choices about priorities for the allocation of internal R&D funding will continue and firms will increasingly depend on universities and other external sources for technological breakthroughs; thus, the need for quality leadership will only grow.

There is today a great opportunity to develop technology for the benefit of the entire world. While local confrontations will continue, the risk of global nuclear warfare has diminished significantly. Among the industrialized nations of the world, there are adequate resources for basic research and the development of technological innovations that benefit all. However, these resources are not unlimited, and we must set priorities and focus on those areas that offer the greatest benefit to society. It would be tragic if we failed to take advantage of the rapidly expanding opportunities in information technology, biotechnology, and other areas that may benefit the global population.

Notes

1. The high cost of capital during the early and middle 1980s, and the focus of many institutional stockholders on short-run indicators of profitability may have put special pressure on corporate research, which has a potentially substantial payoff only in the distant future. While there were some signs that these pressures were diminishing in the early 1990s, they were strongly evident in the late 1980s.

2. For an excellent overview of that system, see David Mowery and Nathan Rosenberg (1993).

3. The evidence of the disappearance of long-run, pioneering research is most vivid in the electronics and information industries. People in the chemical products industries say it is disappearing more slowly there, but it is happening. Pharmaceuticals and biotechnology may be exceptions, in large part because so much of research in these industries is tied to the creation and development of new products.

4. This, of course, is only one of the functions of the research organization. The science-based corporation depends on its research laboratories for more than new technology; strategic technical intelligence and operating "know-how" help sustain competitiveness; the recruitment of trained and talented personnel feeds many corporate needs. As research activities are cut back and redirected, can these functions be sustained? If not, what will be the consequences?

5. In some cases, the CEO may provide that leadership directly. David Sarnoff defined the challenges that drove RCA Laboratories to a series of stunning innovations.

6. The dilemmas arising from this are identified by Leslie Cauley (1995). In the current climate for R&D, they may have difficulty locating a buyer. The David Sarnoff Laboratories in Princeton, New Jersey, once a center for pioneering work in video, liquid crystals, lasers, and other fields, was given away by General Electric in the late 1980s. Its new owner, SRI International, has transformed it into a contract research laboratory.

References

Cauley, Leslie. 1995. "Baby Bells Find It Hard to Put Price on Bellcore." *The Wall Street Journal*, 22 May, 31.

Jensen, Michael C. 1993. "The Modern Industrial Revolution, Exit, and the Failure of Internal Control Systems." *Journal of Finance* 48, no. 3: 831–80.

Mowery, David C., and Nathan Rosenberg. 1993. "The U.S. National Innovation System." In *National Innovation Systems: A Comparative Analysis,* edited by Richard R. Nelson. New York: Oxford University Press.

National Academy of Sciences, Committee on Science, Engineering, and Public Policy. 1993. *Science, Technology, and the Federal Government: National Goals for a New Era.* Washington, D.C.: National Academy of Sciences.

Smith, Hendrich. 1995. *Rethinking America.* New York: Random House.

Spencer, William J. 1990. "Research to Product: A Major U.S. Challenge." *California Management Review* 32, no. 2: 45–53.

Spencer, William J., and Peter Grindley. 1993. "SEMATECH after Five Years: High-Technology Consortia and U.S. Competitiveness." *California Management Review* 35: 9–32.

REFERENCES

Adam, John A. 1990. "Federal Laboratories Meet the Marketplace." *IEEE Spectrum* 28 (October): 39–44.

Alic, John A., et al. 1992. *Beyond Spinoff: Military and Commercial Technologies in a Changing World*. Boston: Harvard Business School Press.

Allison, David K. 1981. *New Eye for the Navy: The Origins of Radar at the Naval Research Laboratory*. Washington, D.C.: U.S. Government Printing Office.

Andrews, Edmund L. 1993. "Swords to Plowshares: The Bureaucratic Snags." *New York Times*, 16 February.

Anthony, Robert N. 1952. *Management Controls in Industrial Research Organizations*. Boston: Graduate School of Business Administration.

Bacon, Glenn, Sara Beckman, David Mowery, and Edith Wilson. 1994. "Managing Product Definition in High Technology Industries: A Pilot Study." *California Management Review* 36: 32–56.

Bartlett, Howard R. 1941. "The Development of Industrial Research in the United States." In *Research—A National Resource*. Report of the National Research Council to the National Resources Planning Board, December 1940. Vol. 2, 19–77. Washington, D.C.: U.S. Government Printing Office.

Baxter, James P. 1968. Reprint. *Scientists Against Time*. Cambridge: MIT Press. Original edition, Boston: Little Brown, 1946.

Beard, Edmund. 1976. *Developing the ICBM: A Study in Bureaucratic Politics*. New York: Columbia University Press.

Beer, John J. 1958. "Coal Tar Dye Manufacture and the Origins of the Modern Industrial Research Laboratory." *Isis* 49: 123–31.

————. 1959. *The Emergence of the German Dye Industry*. Urbana: University of Illinois Press.

Birnbaum, Joel. 1992. Interview by Dorothy Leonard-Barton, November.

Birr, Kendall. 1966. "Science in American Industry." In *Science and Society in the United States*, edited by David D. Van Tassel and Michael G. Hall. Homewood, Ill.: Dorsey Press.

————. 1979. "Industrial Research Laboratories." In *The Sciences in the American Context: New Perspectives*, edited by Nathan Reingold. Washington, D.C.: Smithsonian Institution Press.

Bloch, Erich. 1986. "Basic Research and Economic Health: The Coming Challenge." *Science* 232: 595–99.

Bloch, Erich, and James D. Meindl. 1983. "Some Perspectives from the Field." In *University-Industry Research Relationships: Selected Studies*, edited by

National Science Board, National Science Foundation. Washington, D.C.: U.S. Government Printing Office.

Bolton, E. 1945. Records of E. I. du Pont de Nemours & Co. Series II, part 2, Box 832, Hagley Museum and Library, Wilmington, Delaware.

Booz-Allen and Hamilton, Inc. 1982. *New Product Management for the 1980's.* New York: Booz-Allen & Hamilton, Inc.

Bowen, Kent, Kim Clark, Charles Holloway, and Steven Wheelwright, eds. 1994. *The Perpetual Enterprise Machine: Seven Keys to Corporate Renewal through Successful Product and Process Development.* New York: Oxford University Press.

Bower, Joseph, and Thomas M. Hout. 1988. "Fast-Cycle Capability for Competitive Power." *Harvard Business Review* (November–December): 110–18.

Breech, E. 1949. Research 1949, Accession 65–71, Box 46, Ford Industrial Archives, Ford Motor Company, Dearborn, Michigan.

Brittain, James E. 1971. "The Introduction of the Loading Coil: George A. Campbell and Michael I. Pupin." *Technology and Culture* 11: 36–57.

———. 1976. "C. P. Steinmetz and E. F. W. Alexanderson: Creative Engineering in a Corporate Setting." *IEEE Proceedings* 64: 1413–17.

Bromberg, Joan Lisa. 1983. *Fusion: Science, Politics, and the Invention of a New Energy Source.* Cambridge, Mass.: MIT Press.

———. 1991. *The Laser in America, 1950–1970.* Cambridge: MIT Press.

Brooks, Harvey. 1986. "National Science Policy and Technological Innovation." In *The Positive Sum Strategy: Harnessing Technology for Economic Growth,* edited by Ralph Landau and Nathan Rosenberg. Washington, D.C.: National Academy Press.

Brown, John Seely. 1991. "Research That Reinvents the Corporation." *Harvard Business Review* (January–February): 102–11.

Brown, John Seely, and Susan E. Newman. 1985. "Issues in Cognitive and Social Ergonomics: From Our House to Bauhaus." *Human-Computer Interaction* 1: 359–91.

Browne, Malcolm W. 1994. "Cold War's End Clouds Research as Openings in Science Dwindle." *New York Times,* 20 February. 1:1.

Burgelman, Robert A. 1983. "A Process Model of Internal Corporate Venturing in the Diversified Major Firm." *Administrative Science Quarterly* 28: 223–44.

Bush, Vannevar. 1945. *Science—the Endless Frontier: A Report to the President on a Program for Postwar Scientific Research.* Washington, D.C.: United States Government Printing Office.

Cahan, David. 1982. "Werner Siemens and the Origin of the Physikalisch-Technische Reichsanstalt, 1872–1887." *Historical Studies in the Physical Sciences* 12: 253–83.

————. 1988. *An Institute for an Empire: The Physikalisch-Technische Reichsanstalt, 1871–1918.* Cambridge: Cambridge University Press.

Calantone, R., and Robert G. Cooper. 1979. "A Discriminant Model for Identifying Scenarios of Industrial New Product Failure." *Journal of the Academy of Marketing Science* 7: 163–83.

Carlson, W. Bernard. 1991a. "Building Thomas Edison's Laboratory at West Orange, New Jersey." *History of Technology* 13: 150–67.

————. 1991b. *Innovation as a Social Process: Elihu Thomson and the Rise of General Electric, 1870–1900.* Cambridge: Cambridge University Press.

Carothers, Wallace H. 1929. "An Introduction to the General Theory of Condensation Polymers." *Journal of the American Chemical Society* 51: 2548–559.

————. 1931. "Polymerization." *Chemical Reviews* 8: 353–426.

————. 1932. "Fundamental Research in Organic Chemistry at the Experimental Station—A Review." Records of E. I. du Pont de Nemours & Co., Central Research and Development Department. Accession 1784, Box 16, Hagley Museum and Library, Wilmington, Delaware.

Casson, Mark, ed. 1991. *Global Research Strategy and International Competitiveness.* Cambridge, Mass.: B. Blackwell.

Cauley, Leslie. 1995. "Baby Bells Find It Hard to Put Price on Bellcore." *The Wall Street Journal,* 22 May, B1.

Chandler, Alfred D., Jr. 1962. *Strategy and Structure.* Cambridge, Mass.: MIT Press.

————. 1977. *The Visible Hand: The Managerial Revolution in American Business.* Cambridge, Mass.: Harvard University Press.

————. 1990. *Scale and Scope: The Dynamics of Industrial Capitalism.* Cambridge, Mass.: Belknap Press.

Christensen, Clayton. 1992. "The Innovator's Challenge: Understanding the Influence of Market Environment on Processes of Technology Development in the Rigid Disk Drive Industry." Ph.D. diss., Harvard Business School.

Clark, Kim B. 1989. "What Technology Can Do for Strategy." *Harvard Business Review* (November–December): 94–8.

Clayton, James L., ed. 1970. *The Economic Impact of the Cold War.* New York: Harcourt, Brace & World.

Clinton, William J., and Albert Gore, Jr. 1993. *Technology for America's Economic Growth: A New Direction to Build Economic Strength.* 22 February. Washington, D.C.: U.S. Government Printing Office.

Cohen, Wesley M., and Daniel A. Levinthal. 1990. "Absorptive Capacity: A New Perspective on Learning and Innovation." *Administrative Sciences Quarterly* 35: 128–52.

Cohen, Wesley, Richard Florida, and Richard Goe. 1994. *University-Industry Research Centers in the United States.* Pittsburgh: Carnegie Mellon University.

Cole, Bernard. 1992. "DOE Labs: Models for Tech Transfer." *IEEE Spectrum* 29 (December): 53–68.

Conant, James B. Papers, Harvard University Archives. Cambridge, Massachusetts.

Condon, Edward U. 1942. "Physics in Industry." *Science* 96: 172–74.

Cooper, Robert G. 1975. "Why Industrial New Products Fail." *Industrial Marketing Management* 4: 315–26.

————. 1986. *Winning at New Products.* Reading, Mass.: Addison-Wesley.

Cooper, Robert G., and Elko J. Kleinschmidt. 1986. "An Investigation into the New Product Process: Steps, Deficiencies, and Impact." *Journal of Product Innovation Management* 3: 71–85.

————. 1990. "New Product Success Factors: A Comparison of 'Kills' versus Successes and Failures." *R&D Management* 20, no. 1: 47–63.

Cordes, Colleen. 1994. "The Highest Priority." *The Chronicle of Higher Education* (10 August): A21–A22.

Council on Competitiveness. 1992. *Industry as a Customer of the Federal Laboratories.* Washington, D.C.: Council on Competitiveness.

Cowan, Alison Leigh. 1993. "Unclear Future Forced Board's Hand." *New York Times,* 7 August, 1, 37.

Cusumano, Michael, Yorgis Mylonadis, and Richard Rosenbloom. 1992. "Strategic Maneuvering and Mass Market Dynamics: The Triumph of Beta over VHS." *Business History Review* 66: 51–94.

Dalton, Donald, and Manuel Serapio. 1993. *U.S. Research Facilities of Foreign Companies.* Washington, D.C.: Japan Technology Program, Technology Administration, U.S. Department of Commerce.

David, Burton H., and William P. Hellinger, eds. 1983. *Heterogeneous Catalysis: Selected American Histories.* ACS Symposium Series, 222. Washington, D.C.: American Chemical Society.

David, Edward R., Jr. 1994. "Science in the Post-Cold War Era." *The Bridge* 24 (Spring): 3–8.

David, Paul A., David C. Mowery, and W. Edward Steinmueller. 1992. "Analyzing the Economic Payoffs to Basic Research." *Economics of Innovation and New Technology* 2: 73–90.

Davis, E. W. 1964. *Pioneering with Taconite.* St. Paul: Minnesota Historical Society.

Davis, Lance E., and Daniel J. Kevles. 1974. "The National Research Fund: A Case Study in the Industrial Support of Academic Science." *Minerva* 12: 213–20.

Davis, Stanley M. 1987. *Future Perfect*. Reading, Mass.: Addison-Wesley.

Davisson C. J., and L. Germer. 1927. "Diffraction of Electrons by a Crystal of Nickel." *Physical Review* 30: 705–40.

Desseaur, John H. 1971. *My Years with Xerox: The Billions Nobody Wanted*. Garden City, New York: Doubleday.

Devorkin, David H. 1992. *Science with a Vengeance: How the Military Created U.S. Space Sciences after World War II*. New York: Springer-Verlag.

Dioso, John, and David Hunter. 1992. "Western Firms See Japanese R&D as Key to Success." *Chemical Week* 150, no. 20: 20.

Dougherty, Deborah, and Trudy Heller. 1994. "The Illegitimacy of Successful Product Innovation in Established Firms." *Organization Science* (May): 200–18.

Doyle, John. 1985. "Commentary: Managing New Product Development: How Japanese Companies Learn and Unlearn." In *The Uneasy Alliance: Managing the Productivity-Technology Dilemma*, edited by Kim Clark, Robert Hayes, and Christopher Lorenz. Boston, Mass.: Harvard Business School Press.

Dumaine, Brian. 1989. "How Managers Can Succeed Through Speed." *Fortune* 119 (13 February): 54–74.

Duncan, Robert Kennedy. 1907. *The Chemistry of Commerce*. New York: Harper and Bros.

Dunning, John. 1994. "Multinational Enterprises and Globalization of Innovatory Capacity." *Research Policy* 23: 67–88.

Dyer, Davis. 1993. "Necessity as the Mother of Convention: Developing the ICBM, 1954–1958." Paper presented at the Business History Conference, 19 March, at Boston, Massachusetts.

Erker, Paul. 1990. "Die Verwissenschaftlichung der Industrie: Zur Geschichte der Industrieforschung in den Europäischen und Americkanischen Electrokonzernen 1890–1930." *Zeitschrift für Unternehmensgeschichte* 21: 73–94.

Etzkowitz, Henry. 1988. "The Making of Entrepreneurial University: The Traffic among MIT and the Industry and the Military, 1860–1960." In *Science, Technology, and the Military*. Vol. 2, edited by Everett Mendelsohn, et al. Dordecht, The Netherlands: Kluwer.

Evan, William M., and Paul Olk. 1990. "R&D Consortia: A New Organizational Form." *Sloan Management Review* 31: 37–46.

Fagen, M. D., ed. 1978. *A History of Engineering and Science in the Bell System: National Service in War and Peace (1925–1975)*. N.p.: Bell Telephone Laboratories, Inc.

Feller, I. 1990. "Universities as Engines of R&D-Based Economic Growth: They Think They Can." *Research Policy* 19: 335–48.

Finlay, Douglas, 1991. "Office Printers: Higher Quality, Lower Prices," *Office* 113, no. 5: 42–44.

Fleming, A. P. M. 1917. *Industrial Research in the United States.* London: HM Stationery Office.

Flink, James J. 1988. *The Automobile Age.* Cambridge: MIT Press.

Florida, Richard, and Martin Kenney. 1994. "The Globalization of Innovation: The Economic Geography of Japanese R&D Investment in the United States." *Economic Geography* 70: 344–69.

Ford Industrial Archives. 1951. Press release from Ford news bureau, October 4. AK-74-18056. Ford Motor Company. Dearborn, Michigan.

Forman, Paul. 1987. "Behind Quantum Electronics: National Security as [the] Basis for Physical Research in the United States." Part 1. *Historical Studies in the Physical and Biological Sciences* 19: 149–229.

Foster, Richard. 1982. "Organize for Technology Transfer." *Business Week* 24 May): 24–33.

Freeze, Karen, and Dorothy Leonard-Barton. 1991a. *GE Plastics: Selecting a Partner.* Boston, Mass.: Design Management Institute.

————. 1991b. *Polymer Solutions: Tempest About a Teapot.* Boston, Mass.: Design Management Institute.

Friedel, Robert, and Paul Israel. 1986. *Edison's Electric Light: Biography of an Invention.* New Brunswick: Rutgers University Press.

Furman, Necah Stewart. 1990. *Sandia National Laboratories: The Postwar Decade.* Albuquerque: University of New Mexico Press.

Furnas, C. C. ed. 1948. *Research in Industry: Its Organization and Management.* New York: D. Van Nostrand Company.

Fusfeld, Herbert I., and Carmela S. Haklisch. 1985. "Cooperative R&D for Competitors." *Harvard Business Review* (November–December): 4–11.

Galambos, Louis. 1979. "The American Economy and the Reorganization of the Sources of Knowledge." In *The Organization of Knowledge in Modern America, 1860–1920,* edited by Alexandra Oleson and John Voss. Baltimore: Johns Hopkins University Press.

————. 1992. "Theodore N. Vail and the Role of Innovation in the Modern Bell System." *Business History Review* 66 (Spring): 95–126.

Galison, Peter. 1988. "Physics between War and Peace." In *Science, Technology, and the Military,* edited by Everett Mendelsohn, et al. Dordecht, The Netherlands: Kluwer.

Galison, Peter, and Bruce Hevly. 1992. *Big Science: The Growth of Large-Scale Research.* Stanford: Stanford University Press.

Gambardella, A. 1992. "Competitive Advantages from In-House Scientific Research: The U.S. Pharmaceutical Industry in the 1980s." *Research Policy* 21: 391–407.

Garfield, E. 1993. "Citation Index for Scientific Information." *Science Watch* 4, no. 2: 8.

Geiger, Roger L. 1986. *To Advance Knowledge: The Growth of American Research Universities, 1900–1940.* New York: Oxford University Press.

———. 1992. "Science, Universities, and National Defense, 1945–1970." *Osiris,* 2d ser., 7: 26–48.

———. 1993. *Research and Relevant Knowledge: American Research Universities Since World War II.* New York: Oxford University Press.

Geppert, Linda. 1994. "Industrial R&D: The New Priorities." *IEEE Spectrum* 31 (September): 30–41.

Ghemawat, P., M. E. Porter, and R. A. Rawlinson. 1986. "Patterns of International Coalition Activity." In *Competition in Global Industries,* edited by Michael E. Porter. Boston, Mass.: Harvard Business School Press.

Gibbons, M., and C. Johnson. 1970. "Relationship Between Science and Technology." *Nature* 227 (11 July): 125–27.

Gibson, David V., and Everett M. Rogers. 1994. *R&D Collaboration on Trial: The Microelectronics and Computer Corporation.* Boston, Mass.: Harvard Business School Press.

Glantz, Stanton A., and Norman V. Albers. 1974. "Department of Defense R&D in the University." *Science* 186: 706–11.

Gomes-Casseres, Benjamin. 1988. "Joint Venture Cycles: The Evolution of Ownership Strategies of U.S. MNEs, 1945–75." In *Cooperative Strategies in International Business,* edited by F. J. Contractor and P. Lorange. Lexington, Mass.: Lexington Books.

Gomory, Ralph. 1992. "The Technology-Product Relationship: Early and Late Stages." In *Technology and the Wealth of Nations,* edited by Nathan Rosenberg, Ralph Landau, and David Mowery. Stanford: Stanford University Press.

Goodstein, Judith R. 1991. *Millikan's School: A History of The California Institute of Technology.* New York: W. W. Norton.

Gorn, Michael H. 1988. *Harnessing the Genie: Science and Technology Forecasting for the Air Force, 1944–1986.* Washington, D.C.: Office of Air Force History.

Government-University-Industry Research Roundtable. 1991. "Industrial Perspectives on Innovation and Interactions with Universities" (February). Washington, D.C.: National Academy Press.

Graham, Margaret B. W. 1985a. "Corporate Research and Development: The Latest Transformation." *Technology in Society* 7: 179–95.

————. 1985b. "Industrial Research in the Age of Big Science." *Research on Technological Innovation, Management, and Policy* 2: 47–79.

————. 1986. *RCA & The VideoDisc: The Business of Research*. New York: Cambridge University Press.

Graham, Margaret B. W., and Bettye H. Pruitt. 1990. *R&D for Industry: A Century of Technical Innovation at Alcoa*. New York: Cambridge University Press.

Graham, Otis L., Jr. 1992. *Losing Time: The Industrial Policy Debate*. Cambridge: Harvard University Press.

Granstrand, Ove, Lars Håkanson, and Sören Sjölander, eds. 1992. *Technology Management and International Business*. New York: John Wiley & Sons.

Greenberg, Daniel S. 1966. "Basic Research: The Political Tides are Shifting." *Science* 152: 1724–26.

Greenewalt, Crawford H. Manhattan Project Diaries, Accession 1889, Hagley Museum and Library, Wilmington, Delaware.

————. Papers. Accession 1814, Box 37, Hagley Museum and Library, Wilmington, Delaware.

Griliches, Zvi. 1986. "Productivity, R&D and Basic Research at the Firm Level in the 1970s." *American Economic Review* 76, no. 1: 141–54.

————. 1992. "The Search for R&D Spillovers." *The Scandinavian Journal of Economics* 94: 29–47.

————. 1994. "Productivity, R&D and the Data Constraint." *American Economic Review* 84, no. 1: 1–23.

Grindley, P. 1990. "Winning Standards Contests: Using Product Standards in Business Strategy." *Business Strategy Review* (Spring): 71–84.

Grindley, P., D. C. Mowery, B. Silverman. 1994. "Sematech and Collaborative Research: Lessons in the Design of High-Technology Consortia." *Journal of Policy Analysis and Management* 13: 723–58.

Grossman, Gene, and Elhanan Helpman. 1991. *Innovation and Growth in the Global Economy*. Cambridge, Mass.: MIT Press.

Gupta, Ashok, and David Wilemon. 1990. "Accelerating the Development of Technology-Based New Products." *California Management Review* (Winter): 24–44.

Hall, Bronwyn. 1993. "New Evidence on the Impacts of R&D." *Brookings Papers on Economic Activity: Microeconomics*. Washington, D.C.: Brookings Institution.

Hamel, Gary, Yves Doz, and C. K. Prahalad. 1989. "Collaborate with Your Competitors—and Win." *Harvard Business Review* (January–February): 133–39.

Hammer, Michael, and James Champy. 1993. *Reengineering the Corporation: A Manifesto for Business Revolution*. New York: Harper Collins.

Harrigan, Kathryn R. 1984. "Joint Ventures and Competitive Strategy." Working paper, Graduate School of Business, Columbia University, New York.

Hauser, John R., and Don Clausing. 1988. "The House of Quality." *Harvard Business Review* (May–June): 63–73.

Hayes, Robert. 1985. "Strategic Planning—Forward in Reverse?" *Harvard Business Review* (November–December): 111–19.

Haynes, Williams. 1945. *The American Chemical Industry*. Vol. 2. New York: Van Nostrand.

Herbert, Evan. 1989. "Japanese R&D in the United States." *Research Technology Management* 32 (November–December): 11–20.

Hercules, D. M., and J. W. Enyart. 1983. "Report on the Questionnaire on Current Exchange Programs Between Industries and Universities." Council on Chemical Research, University-Industry Interaction Committee.

Hershberg, James. 1993. *James B. Conant: Harvard to Hiroshima and the Making of the Nuclear Age*. New York: Knopf.

Hewlett, Richard G. 1976. "Beginnings of Development in Nuclear Technology." *Technology and Culture* 17: 465–78.

Hewlett, Richard G., and Oscar E. Anderson, Jr. 1962. *The New World: A History of the United States Atomic Energy Commission, 1939–1946*. University Park: Pennsylvania State University Press.

Hewlett, Richard G., and Francis Duncan. 1972. *Atomic Shield: A History of the United States Atomic Energy Commission*. Washington, D.C.: U.S. Atomic Energy Commission.

———. 1974. *The Nuclear Navy, 1946–1962*. Chicago: University of Chicago Press.

Hewlett, Richard G., and Jack M. Holl. 1989. *Atoms for Peace and War, 1953–1961: Eisenhower and the Atomic Energy Commission*. Berkeley: University of California Press.

Hindle, Brooke. 1956. *The Pursuit of Science in Revolutionary America*. Chapel Hill: University of North Carolina Press.

Hirsh, Richard. 1989. *Technology and Transformation in the Electric Utility Industry*. New York: Cambridge University Press.

Hitt, Michael, and R. Duane Ireland. 1985. "Corporate Distinctive Competence, Strategy, Industry, and Performance." *Strategic Management Journal* 6: 273–93.

Hladik, Karen. 1985. *International Joint Ventures*. Lexington, Mass.: D.C. Heath.

Hoch, Paul K. 1988. "The Crystallization of a Strategic Alliance: The American Physics Elite and the Military in the 1940s." In *Science, Technology, and the*

Military. Vol. 1, edited by Everett Mendelsohn et al. Dordecht, The Netherlands: Kluwer.

Hof, Robert D. 1993. "Hewlett-Packard Digs Deep for a Digital Future." *Business Week* (18 October): 72–75.

Hoover, Herbert. 1926. "The Vital Need for Greater Financial Support to Pure Science Research." *Mechanical Engineering* 48 (January): 6–8.

Hopkins, D. S., and E. L. Bailey. 1971. "New Product Pressures." *Conference Board Record* 8: 16–24.

Hounshell, David A. 1980. "Edison and the Pure Science Ideal in 19th-Century America." *Science* 207: 612–17.

———. 1989. "The Modernity of Menlo Park." In *Working at Inventing*, edited by William S. Pretzer. Dearborn, Mich.: Henry Ford Museum and Greenfield Village.

———. 1992. "Du Pont and the Management of Large-Scale Research and Development." In *Big Science: The Growth of Large-Scale Research*, edited by Peter Galison and Bruce Hevly. Stanford: Stanford University Press.

Hounshell, David A., and John Kenly Smith, Jr. 1988. *Science and Corporate Strategy: Du Pont R&D, 1902–1980*. Cambridge: Cambridge University Press.

Howard, Robert. 1992. "The CEO as Organizational Architect: An Interview with Xerox's Paul Allaire." *Harvard Business Review* (September–October): 107–21.

Hughes, Thomas P. 1971. *Elmer Sperry: Inventor and Engineer.* Baltimore: The Johns Hopkins University Press.

Illinois Institute of Technology Research Institute. 1968. *Technology in Retrospect and Critical Events in Science.* Chicago: Illinois Institute of Technology.

Industrial Research Institute. 1994. *Annual Report—1994.* Washington, D.C.: Industrial Research Institute.

Israel, Paul, 1989. "Telegraphy and Edison's Invention Factory." In *Working at Inventing: Thomas A. Edison and the Menlo Park Experience*, edited by William S. Pretzer. Dearborn, Mich.: Henry Ford Museum and Greenfield Village.

———. 1992. *From Machine Shop to Inventor.* Baltimore: The Johns Hopkins University Press.

Jacobson, G., and J. Hillkirk. 1986. *Xerox: American Samurai.* New York: Macmillian.

Jaffe, Adam. 1986. "Technological Opportunity and Spillovers of R&D." *American Economic Review* 76: 984–1001.

———. 1989. "Real Effects of Academic Research." *American Economic Review* 79: 957–70.

Jaffe, Adam, Manuel Trajtenberg, and Rebecca Henderson. 1993. "Geographic Location of Knowledge Spillovers as Evidenced by Patent Citations." *Quarterly Journal of Economics* (August): 577–98.

Jenkins, Reese V. 1975. *Images and Enterprise: Technology and the American Photographic Industry, 1839–1925.* Baltimore: The Johns Hopkins University Press.

Jensen, Michael C. 1993. "The Modern Industrial Revolution, Exit, and the Failure of Internal Control Systems." *Journal of Finance* 48, no. 3: 831–80.

Jewett, Frank B. 1947. "The Future of Scientific Research in the Postwar Era." In *Science in Progress,* no. 5, edited by George A. Baitsell. New Haven: Yale University Press.

Johnson, Harry G. 1965. "Federal Support of Basic Research: Some Economic Issues." In *Basic Research and National Goals: A Report to the Committee on Science and Astronautics of the U.S. House of Representatives.* Washington, D.C.: National Academy of Sciences.

Johnson, Jeffrey A. 1990a. "Academic Proletarian, . . . Professional? Shaping Professionalization for German Industrial Chemists, 1887–1920." In *German Professions, 1800–1950,* edited by Geoffrey Cocks and Konrad H. Jarausch. New York: Oxford University Press.

———. 1990b. *The Kaiser's Chemists: Science and Modernization in Imperial Germany.* Chapel Hill: University of North Carolina Press.

Johnstone, Robert. 1992. "Research: Setting Up on Enemy Ground." *Far Eastern Economic Review* 155, no. 24: 54–55.

Jones, Daniel P. 1969. "The Role of Chemists in War Gas Research in the United States during World War I." Ph.D. diss., University of Wisconsin.

Jones, Kenneth MacDonald. 1975. "Science, Scientists, and Americans: Images of Science and the Formation of Federal Science Policy, 1945–1950." Ph.D. diss., Cornell University.

Kahn, E. J. 1986. *The Problem Solvers: A History of Arthur D. Little, Inc.* Boston: Little, Brown.

Kanter, Rosabeth Moss. 1988. "When a Thousand Flowers Bloom: Structural, Collective, and Social Conditions for Innovation in Organizations." In *Research in Organizational Behavior,* edited by Barry M. Staw and L. L. Cummings. Vol. 10. Greenwich, Conn.: JAI Press.

Kantrow, Alan. 1986. "Wide-Open Management at Chaparral Steel: An Interview with Gordon E. Forward." *Harvard Business Review* (May–June): 96–102.

Kargon, Robert H. 1982. *The Rise of Robert Millikan: Portrait of a Life in American Science.* Ithaca, New York: Cornell University Press.

Kay, Herbert. 1965. "Harnessing the R&D Monster." *Fortune* (January): 160–163, 196–198.

Kearns, David T., and David A. Nadler. 1992. *Prophets in the Dark: How Xerox Reinvented Itself and Beat Back the Japanese.* New York: Harper Collins.

Kevles, Daniel, J. 1977. "The National Science Foundation and the Debate over Postwar Research Policy, 1942–1945: A Political Interpretation of *Science—The Endless Frontier.*" *Isis* 68: 5–26.

———. 1978. *The Physicists: The History of a Scientific Community in Modern America.* New York: Knopf.

———. 1987. "The Remilitarization of American Science: Historical Reflections." Manuscript.

———. 1988. "An Analytical Look at R&D and the Arms Race." In *Science, Technology, and the Military.* Vol. 2, edited by Everett Mendelsohn, et al. Dordecht, The Netherlands: Kluwer.

———. 1990. "Principles and Politics in Federal R&D Policy, 1945–1990: An Appreciation of the Bush Report." Introduction to *Science—The Endless Frontier* by Vannevar Bush. Reprint edition. Washington, D.C.: National Science Foundation.

Khazam, J., and D. C. Mowery. 1994. "The Commercialization of RISC: Strategies for the Creation of Dominant Designs." *Research Policy* 23: 89–102.

Kilborn, Peter T. 1993. "The Ph.D.'s Are Here, But the Lab Isn't Hiring," *New York Times,* 18 July, E3.

Kline, Ronald. 1986. "The Origins of Industrial Research at the Westinghouse Electric Company, 1886–1922." Paper presented at the annual meeting of Society for the History of Technology, 25 October, Pittsburgh.

———. 1987. "R&D: Organizing for War." *IEEE Spectrum* 24 (November): 54–60.

———. 1992. *Steinmetz: Engineer and Socialist.* Baltimore: The Johns Hopkins University Press.

Kline, Stephen J., and Nathan Rosenberg. 1986. "An Overview of Innovation." In *The Positive Sum Strategy,* edited by Ralph Landau and Nathan Rosenberg. Washington, D.C.: National Academy Press.

Knoedler, Janet T. 1991. "Backward Linkages to Industrial Research in Steel, 1870–1930." Ph.D. diss., University of Tennessee.

———. 1993a. "Early Examples of User-Based Industrial Research." *Business and Economic History* 22 (Fall): 285–94.

———. 1993b. "Market Structure, Industrial Research, and Consumers of Innovation: Forging Backward Linkages to Research in the Turn-of-the-Century U.S. Steel Industry." *Business History Review* 67 (Spring): 98–139.

Kohlstedt, Sally Gregory. 1976. *The Formation of the American Scientific Community.* Urbana: University of Illinois Press.

Komons, Nick A. 1966. *Science and the Air Force: A History of the Air Force Office of Scientific Research.* Arlington, Virginia: Office of Aerospace Research.

Koppes, Clayton R. 1982. *JPL and the American Space Program: A History of the Jet Propulsion Laboratory.* New Haven: Yale University Press.

Lamar, Howard R. 1969. "Frederick Jackson Turner." In *Pastmasters,* edited by Marcus Cunliffe and Robin Winks. New York: Harper & Row.

Langrish, J., M. Gibbons, C. Johnson, and F. R. Jevons. 1972. *Wealth from Knowledge.* New York: Halsted/John Wiley.

Lasby, Clarence G. 1971. *Project Paperclip: German Scientists and the Cold War.* New York: Atheneum.

Lazonick, William, William Mass, and Jonathan West. 1995. "Strategy, Structure, and Performance: Comparative-Historical Foundations of the Theory of Competitive Advantage." In *Proceedings of the Conference on Business History,* October 24 and 25, 1994, Rotterdam, The Netherlands, edited by Mila Davids, Ferry de Goey, and Dirk de Wit. Rotterdam: Centre of Business History, Erasmus University.

Leonard-Barton, Dorothy. 1987. "The Case for Integrative Innovation: An Expert System at Digital." *Sloan Management Review* 29, no. 1: 7–19.

———. 1989. "Implementing New Production Technologies: Exercises in Corporate Learning." In *Managing Complexity in High Technology Industries: Systems and People,* edited by Mary Von Glinow and Susan Mohrman. New York: Oxford University Press.

———. 1992a. "Core Capabilities and Core Rigidities: A Paradox in Managing New Product Development." *Strategic Management Journal* 13: 111–25.

———. 1992b. "The Factory as a Learning Laboratory." *Sloan Management Review* 34, no. 1: 23–26.

———. 1992c. "Inanimate Integrators: A Block of Wood Speaks." *Design Management Journal* 2, no. 3: 61–67.

———. 1995. *Wellsprings of Knowledge: Building and Sustaining the Sources of Innovation.* Boston: Harvard Business School Press.

Leonard-Barton, Dorothy, and Deepak Sinha. 1990. "Dependency, Involvement, and User Satisfaction: The Case of Internal Software Development." Working paper #91-008, Harvard Business School, Cambridge, Mass.

Lesher, Richard L. 1963. "Independent Research Institutes and Industrial Application of Aerospace Research." Ph.D. diss., Indiana University.

Leslie, Stuart W. 1983. *Boss Kettering: Wizard of General Motors.* New York: Columbia University Press.

———. 1993. *The Cold War and American Science.* New York: Columbia University Press.

Levine, Alan J. 1994. *The Missile and Space Race.* Greenwood, Conn.: Praeger.

Levine, Arnold S. 1982. *Managing NASA in the Apollo Era.* Washington, D.C.: National Aeronautics and Space Administration.

Levine, David O. 1986. *The American College and the Culture of Aspiration, 1915–1940*. Ithaca: Cornell University Press.

Liebenau, Jonathan. 1987. *Medical Science and Medical Industry: The Formation of the American Pharmaceutical Industry*. Baltimore: The Johns Hopkins University Press.

Link, A. N. 1995. "Research Joint Ventures: Patterns from *Federal Register* Filings." Economics working paper, Center for Applied Research, Bryan School of Business and Economics, University of North Carolina-Greensboro.

Little, Arthur D. 1913. "Industrial Research in America." *Journal of Industrial and Engineering Chemistry* 5: 793–801.

Logsdon, John. 1970. *The Decision to Go to the Moon: Project Apollo and the National Interest*. Cambridge: MIT Press.

Maidique, Modesto A., and Billie Jo Zirger. 1985. "The New Product Learning Cycle." *Research Policy* 14 (December): 229–313.

Mansfield, Edwin. 1991. "Academic Research and Industrial Innovation." *Research Policy* 20: 1–12.

McCurdy, Howard E. 1993. *High Technology and Organizational Change in the American Space Program*. Baltimore: Johns Hopkins University Press.

McDougall, Walter. 1985. *The Heavens and the Earth: A Political History of the Space Age*. New York: Basic Books.

Mees, C. E. Kenneth. 1916. "Organization of Industrial Research Laboratories." *Science* 43: 763–73.

———. 1920. *The Organization of Industrial Scientific Research*, New York: McGraw-Hill.

Melman, Seymour. 1965. *Our Depleted Society*. New York: Holt, Rinehart and Winston.

Merck & Co., Inc. 1991. *Values and Visions: A Merck Century*. Rahway, N.J.: Merck & Co., Inc.

Meyer-Thurow, Georg. 1982. "The Industrialization of Invention: A Case Study from the German Chemical Industry," *Isis* 73: 363–81.

Millard, Andre. 1990. *Edison and the Business of Innovation*. Baltimore: The Johns Hopkins University Press.

Millikan, Robert A. 1919. "The New Opportunity in Science." *Science* 50: 85–297. Quoted in Robert H. Kargon, *The Rise of Robert Millikan: Portrait of a Life in American Science*. Ithaca, New York: Cornell University press.

Moffat, Susan. 1991. "Picking Japan's Research Brains." *Fortune* 123 (25 March): 85–96.

Morin, Alexander J. 1993. *Science Policy and Politics*. Englewood Cliffs, N.J.: Prentice-Hall.

Morone, Joseph G. 1993. *Winning in High-Tech Markets: The Role of General Management.* Boston: Harvard Business School Press.

Morris, Peter J. T. 1989. *The American Synthetic Rubber Research Program.* Philadelphia: University of Pennsylvania Press.

Mowery, David C. 1981. "The Emergence and Growth of Industrial Research in American Manufacturing, 1899–1945," Ph.D. diss., Stanford University.

————. 1983a. "Economic Theory and Government Technology Policy." *Policy Sciences* 16, no. 2: 27–43.

————. 1983b. "The Relationship Between Contractual and Intrafirm Forms of Industrial Research in American Manufacturing, 1900–1944." *Explorations in Economic History* 20 (October): 351–74.

————. 1987. *Alliance Politics and Economics: Multinational Joint Ventures in Commercial Aircraft.* Cambridge, Mass.: Ballinger Publishers.

————. 1989. "Collaborative Ventures between U.S. and Foreign Manufacturing Firms." *Research Policy* 18: 19–32.

Mowery, David C., and Nathan Rosenberg. 1989. *Technology and the Pursuit of Economic Growth.* New York: Cambridge University Press.

————. 1993. "The U.S. National Innovation System." In *National Innovation Systems: A Comparative Analysis,* edited by Richard R. Nelson. New York: Oxford University Press.

Mueller, W. F. 1962. "The Origins of the Basic Inventions Underlying Du Pont's Major Product and Process Inventions, 1920 to 1950." In *The Rate and Direction of Inventive Activity,* edited by Richard R. Nelson. Princeton, N.J.: Princeton University Press.

National Academy of Sciences, Committee on Criteria for Federal Support of Research and Development. 1995. *Allocating Federal Funds for Science and Technology.* Washington, D.C.: National Academy of Sciences.

National Academy of Sciences, Committee on Science, Engineering, and Public Policy. 1993. *Science, Technology, and the Federal Government: National Goals for a New Era.* Washington, D.C.: National Academy of Sciences.

National Resources Planning Board. 1938. *Research—A National Resource.* Vol. 1. Washington, D.C.: U.S. Government Printing Office.

National Science Foundation. 1991. *Academic Science/Engineering: R&D Expenditures, Fiscal Year 1989.* NSF90-321, Detailed Statistical Tables. Washington, D.C.: National Science Foundation:

National Science Foundation. National Science Board. 1991. *Science and Engineering Indicators—1991.* Washington, D.C.: U.S. Government Printing Office.

National Science Foundation. National Science Board. Committee on Industrial Support for R&D. 1992. *The Competitive Strength of U.S. Industrial Science and Technology: Strategic Issues.* Washington, D.C., U.S. Government Printing Office.

National Science Foundation. National Science Board. 1993. *Science and Engineering Indicators—1993.* Washington, D.C.: U.S. Government Printing Office.

Nelson, Richard R. 1959. "The Economics of Invention: A Survey of the Literature." *The Journal of Business* 32 (April): 101–7.

————. 1962. "The Link Between Science and Invention: The Case of the Transistor." In *The Rate and Direction of Inventive Activity: Economic and Social Factors,* edited by Richard R. Nelson. Princeton: Princeton University Press.

————. 1963. "The Impact of Arms Reduction on Research and Development." *American Economic Review* 53: 435–46.

————. 1990. "U.S. Technological Leadership: Where Did It Come From and Where Did It Go?" *Research Policy* 19: 117–32.

————. 1991. "Capitalism as an Engine of Progress." *Research Policy* 20: 193–214.

————, ed. 1993. *National Innovation Systems: A Comparative Analysis.* New York: Oxford University.

Nelson, Richard R., and Richard C. Levin. 1986. "The Influence of Science University Research and Technical Societies on Industrial R&D and Technical Advance." Research Program on Technological Change, Policy discussion paper 3, Yale University, New Haven, Conn.

Neushul, Peter. 1993. "Science, Technology, and the Arsenal of Democracy: Production Research and Development during World War II." Ph.D. diss., University of California, Santa Barbara.

Nevins, Allan, and Frank Ernest Hill. 1962. *Ford: Decline and Rebirth, 1933–1962.* New York: Charles Scribner's Sons.

New York Times. 1992. "University of California Proposes Laboratory-to-Marketplace Link," 11 December, A14.

Noble, David. 1977. *America by Design: Science, Technology, and the Rise of Corporate Capitalism.* New York: Knopf.

Ouchi, William G., and Michele Kremen Bolton. 1988. "The Logic of Joint Research and Development." *California Management Review* (Spring): 9–33.

Owens, Larry. 1994. "The Counter-productive Management of Science in the Second World War: Vannevar Bush and the Office of Scientific Research and Development." *Business History Review* 68 (Winter): 515–76.

Pake, George. 1985. "Research at Xerox PARC: A Founder's Assessment." *IEEE Spectrum* (October): 54–61.

Parascandola, John. 1983. "Charles Holmes Herty and the Effort to Establish an Institute for Drug Research in PostWorld War I America." In *Chemistry and Modern Society,* edited by John Parascandola and James C. Wharton. Washington, D.C.: American Chemical Society.

————. 1985. "Industrial Research Comes of Age: The American Pharmaceutical Industry, 1920–1940." *Pharmacy in History* 27: 12–21.

————. 1990. "The 'Preposterous Provision': The American Society for Pharmacology and Experimental Therapeutics' Ban on Industrial Pharmacologists, 1908–1941." In *Pill Peddlers: Essays on the History of the Pharmaceutical Industry,* edited by Jonathan Liebenau et al. Madison: American Institute of the History of Pharmacy.

————. 1992. *The Development of American Pharmacology: John J. Abel and the Shaping of a Discipline.* Baltimore: The Johns Hopkins University Press.

Patel, Pari, and Keith Pavitt. 1991. "Large Firms in the Production of the World's Technology: An Important Case of 'Non-Globalization.'" *Journal of International Business Studies* 22: 1–21.

Pavitt, Keith. 1991. "What Makes Basic Research Economically Useful?" *Research Policy* 20: 109–19.

Peck, Merton J. 1986. "Joint R&D: The Case of the Microelectronics and Computer Technology Corporation." *Research Policy* 15: 219–32.

Perry, Teklu, and Paul Wallich. 1985. "Inside the PARC: The Information Architect." *IEEE Spectrum* (October): 67–75.

Phillips, S. 1989. "When U.S. Joint Ventures with Japan Go Sour." *Business Week* (24 July): 30–31.

Piore, Michael, and Charles Sabel. 1984. *The Second Industrial Divide: Possibilities for Prosperity.* New York: Basic Books.

Pollack, Andrew. 1993. "Cutbacks at Kodak Lab in Japan Create Unease." *New York Times,* 15 February.

Porter, Michael E., and Mark B. Fuller. 1986. "Coalitions and Global Strategy." In *Competition in Global Industries,* edited by Michael E. Porter. Boston: Harvard Business School Press.

Prahalad, C. K., and Gary Hamel. 1990. "The Core Competence of the Corporation." *Harvard Business Review* (May–June): 79–91.

Price, William J., and Lawrence W. Bass. 1969. "Scientific Research and the Innovative Process." *Science* 164 (16 May): 802–06.

Pugh, Emerson W. 1986. "Research." In *IBM's Early Computers,* edited by Charles J. Bashe, Lyle R. Johnson, John H. Palmer, and Emerson W. Pugh. Cambridge: MIT Press.

Rabi, I. I. 1965. "The Interaction of Science and Technology." In *The Impact of Science on Technology,* edited by Aaron W. Warner, Dean Morse, and Alfred S. Eichner. New York: Columbia University Press.

Rae, John. 1979. "The Application of Science to Industry." In *The Organization of Knowledge in Modern America, 1860–1920,* edited by Alexandra Oleson and John Voss. Baltimore: The Johns Hopkins University Press.

Ramo, Simon. 1988. *The Business of Science: Winning and Losing in the High-Tech Age.* New York: Hill and Wang.

Records of E. I. du Pont de Nemours & Co. Hagley Museum and Library, Wilmington, Delaware.

Redmond, Kent C., and Thomas M. Smith. 1980. *Project Whirlwind: The History of a Pioneer Computer.* Bedford, Mass.: Digital Press.

Reich, Leonard S. 1983. "Irving Langmuir and the Pursuit of Science and Technology in the Corporate Environment." *Technology and Culture* 24: 199–221.

———. 1985. *The Making of American Industrial Research: Science and Business at GE and Bell, 1876–1926.* New York: Cambridge University Press.

Reich, Robert B. 1990. "Who Is Us?" *Harvard Business Review* (January–February): 53–64.

———. 1991. *The Work of Nations: Preparing Ourselves for 21st-Century Capitalism.* New York: Knopf.

Reingold, Nathan. 1972. "American Indifference to Basic Research." In *Nineteenth-Century American Science: A Reappraisal,* edited by George H. Daniels. Evanston, Ill.: Northwestern University Press.

———. 1976. "Definitions and Speculations: The Professionalization of Science in America in the Nineteenth Century." In *The Pursuit of Knowledge in the Early American Republic: American Scientific and Learned Societies from the Colonial Times to the Civil War,* edited by Alexandra Oleson and Sanborn C. Brown. Baltimore: Johns Hopkins University Press.

Report of the President's Committee on the Impact of Defense and Disarmament. 1965. President's Committee on the Impact of Defense and Disarmament. Washington, D.C.: U.S. Government Printing Office.

Rigdon, Joan E., and Joann S. Lublin. 1993. "Kodak Seeks Outsider to be Chairman, CEO." *Wall Street Journal,* 9 August.

Roland, Alex. 1985. *Model Research: The National Advisory Committee for Aeronautics, 1915–1958.* 2 vols. Washington, D.C.: National Aeronautics and Space Administration.

Romer, Paul. 1986. "Increasing Returns and Long-Run Growth." *Journal of Political Economy* 94, no. 5: 1002–037.

Ronstadt, Robert. 1977. *Research and Development Abroad by U.S. Multinationals.* New York: Praeger.

Rosenberg, Nathan. 1982. *Inside the Black Box: Technology and Economics.* New York: Cambridge University Press.

————. 1990. "Why Do Firms Do Basic Research (With Their Own Money)?" *Research Policy* 19: 165–74

Rosenberg, Nathan and Richard R. Nelson. 1994. "American Universities and Technical Advance in Industry." *Research Policy* 23: 323–48.

Rosenberg, Robert. 1983. "American Physics and the Origins of Electrical Engineering," *Physics Today* 36: 48–54.

Rothwell, Roy, et al. 1973. *Project SAPPHO—A Comparative Study of Success and Failure in Industrial Innovation.* London: Centre for the Study of Industrial Innovation.

————. 1974. "SAPPHO Updated—Project SAPPHO Phase II." *Research Policy* 3: 258–91.

Rowland, Henry A. 1902. *Physical Papers of Henry A. Rowland.* Baltimore: The Johns Hopkins University Press.

Rudolph, Barbara. 1991. "Follow That Brain Wave." *Time*, 12 August, 69.

Rumelt, Richard. 1986. Reprint. *Strategy, Structure and Economic Performance.* Harvard Business School Classics. Boston: Harvard Business School Press. Original edition, Boston, Harvard Business School Press, 1974.

Russo, Arturo. 1981. "Fundamental Research at Bell Laboratories: The Discovery of Electron Diffraction." *Historical Studies in the Physical Sciences* 12: 117–60.

Sanderson, Susan Walsh, and Vic Uzumeri. 1990. "Strategies for New Product Development and Renewal: Design-Based Incrementalism." Working paper, Center for Science and Technology Policy, School of Management, Rensselaer Polytechnic Institute.

Sapolsky, Harvey. 1990. *Science and the Navy: The History of the Office of Naval Research.* Princeton: Princeton University Press.

————. 1994. "Financing Science After the Cold War." In *The Fragile Contract: University Science and the Federal Government*, edited by David H. Griton and Kenneth Kenniston. Cambridge: MIT Press.

Schriesheim, Alan. 1990. "Toward a Golden Age for Technology Transfer." *Issues in Science and Technology* 7, no. 2: 52–8.

Schwartz, Peter. 1991. *The Art of the Long View: Planning for the Future in an Uncertain World.* New York: Doubleday Currency.

Science. 1937. "The Westinghouse Research Fellowships." 86 (31 December): 605–06.

Schweber, Silvan S. 1988. "The Mutual Embrace of Science and the Military: ONR and the Growth of Physics in the United States after World War II." In *Science, Technology, and the Military.* Vol. 1., edited by Everett Mendelsohn, et al. Dordecht, The Netherlands: Kluwer.

Scott, John T. 1989. "Historical and Economic Perspectives of the National Cooperative Research Act." In *Cooperative Research and Development*, edited by Albert N. Link and Gregory Tassey. Boston: Kluwer.

Seidel, Robert W. 1983. "Accelerating Science: The Postwar Transformation of the Lawrence Radiation Laboratory." *Historical Studies in the Physical Sciences* 13: 375–400.

———. 1986. "A Home for Big Science: The Atomic Energy Commission's Laboratory System." *Historical Studies in the Physical and Biological Sciences* 16: 137–75.

———. 1987. "From Glow to Flow: A History of Military Laser Research and Development." *Historical Studies in the Physical and Biological Sciences* 18: 111–47.

———. 1990. "Clio and the Complex: Recent Historiography of Science and National Security." *Proceedings of the American Philosophical Society* 134: 420–41.

———. 1994. "Accelerators and National Security: The Evolution of Science Policy for High-Energy Physics, 1947–1967." *History and Technology* 11:361–91.

Serum, James. 1992. Interview by Dorothy Leonard-Barton, November.

Servos, John W. 1990. *Physical Chemistry from Ostwald to Pauling: The Making of a Science in America*. Princeton: Princeton University Press.

———. 1994. "Changing Partners: The Mellon Institute, Private Industry, and the Federal Patron," *Technology and Culture* 35 (April): 221–57.

Shapley, Deborah, and Rustum Roy. 1985. *Lost at the Frontier*. Philadelphia: ISI Press.

Sherwin, Chalmers W., and Raymond S. Isenson. 1967. "Project Hindsight: A Defense Department Study of The Utility of Research." *Science* 156 (June 23): 1571–577.

Shiba, Shoji, Richard Lynch, Ira Moskowitz, and John Sheridan. 1991. "Step by Step KJ Method." CQM Document No. 2, The Center for Quality Management, Analog Devices, Wilmington, Massachusetts.

Shuen, Amy S. 1993. "Co-Developed Know-how Assets in Technology Partnerships." Haas School of Business, University of California, Berkeley.

Sigethy, Robert. 1980. "The Air Force Organization for Basic Research, 1945–1970: A Study in Change." Ph.D. diss., American University.

Simon, Herbert. 1969. *The Sciences of the Artificial*. Cambridge: MIT Press.

Smith, Bruce L. R. 1990. *American Science Policy Since World War II*. Washington, D.C.: Brookings Institution.

Smith, Douglas K., and Robert C. Alexander. 1988. *Fumbling the Future—How Xerox Invented, Then Ignored, the First Personal Computer*. New York: William Morrow.

Smith, Hendrich. 1995. *Rethinking America.* New York: Random House.

Smith, T., W. J. Spencer, T. J. Kessler, and J. Elkind. 1988. "Why New Business Initiatives Fail." Xerox Corporation study, Stamford, Connecticut.

Solo, Robert A. 1962. "Gearing Military R&D to Economic Growth." *Harvard Business Review* (November–December): 49–60.

Souder, William. 1987. *Managing New Product Innovations.* Lexington, Mass.: Lexington Books.

Spencer, William J. 1990. "Research to Product: A Major U.S. Challenge," *California Management Review* 32, no. 2: 45–53.

Spencer, William J., and Peter Grindley. 1993. "SEMATECH after Five Years: High-Technology Consortia and U.S. Competitiveness." *California Management Review* 35: 9–32.

Stalk, George, Jr. 1988. "Time—The Next Source of Competitive Advantage." *Harvard Business Review* 66 (July–August): 41–51.

Stalk, George, Jr., and Thomas M. Hout. 1990. *Competing Against Time.* New York: The Free Press.

Steen, Kathryn. 1995. "Wartime Catalyst and Postwar Reaction: The Making of the U.S. Synthetic Organic Chemicals Industry, 1910–1930." Ph.D. diss., University of Delaware.

Stevenson, Earl Place. 1953. *Scatter Acorns That Oaks May Grow: Arthur D. Little, Inc., 1886–1953.* New York: Newcomen Society of North America.

Stine, Jeffrey. 1986. *History of U.S. Science Policy Since World War II: Report of the Task Force on Science Policy.* U.S. House of Representatives Committee on Science and Technology. Washington, D.C.: U.S. Government Printing Office.

Stuckey, J. S. 1983. *Vertical Integration and Joint Ventures in the Aluminum Industry.* Cambridge, Mass.: Harvard University Press.

Sturchio, Jeffrey L. 1985. "Experimenting with Research: Kenneth Mees, Eastman Kodak, and the Challenges of Diversification." Paper presented at the R&D Pioneers Conference, 7 October, Wilmington, Delaware.

Sturm, Thomas. 1967. *The USAF Scientific Advisory Board: Its First Twenty Years, 1944–1964.* Washington, D.C.: Office of Air Force History, U.S. Air Force.

Swann, John P. 1988. *Academic Scientists and the Pharmaceutical Industry: Cooperative Research in Twentieth-Century America.* Baltimore: The Johns Hopkins University Press.

Sweet, William. 1993. "IBM Cuts Research in Physical Sciences at Yorktown Heights and Almaden." *Physics Today* 46, no. 6: 75–9.

Teece, David J. 1977. *The Multinational Corporation and the Costs of International Technology Transfer.* Cambridge, Mass.: Ballinger.

————. 1988. "Technological Change and the Nature of the Firm." In *Technical Change and Economic Theory*, edited by Giovanni Dosi, C. Freeman, R. Nelson, G. Silverberg, and L. Soete. London: Frances Pinter.

————. 1992. "Competition, Cooperation, and Innovation: Organizational Arrangements for Regimes of Rapid Technological Progress." *Journal of Economic Behavior and Organization* 18, no. 1: 1–25.

Teece, David J., Gary Pisano, and Amy Shuen. 1992. "Dynamic Capabilities and Strategic Management." Haas School of Business, University of California, Berkeley.

Thackray, Arnold. 1983. "University-Industry Connections and Chemical Research: An Historical Perspective." In *University-Industry Research Relationships: Selected Studies*, edited by National Science Board, National Science Foundation. Washington, D.C.: U.S. Government Printing Office, 1983.

Tiffany, Paul A. 1986. "Corporate Culture and Corporate Change: The Origins of Industrial Research at the United States Steel Corporation, 1901–1929." Paper presented at the Annual Meeting of the Society for the History of Technology, October 25, Pittsburgh.

Tobey, Ronald C. 1971. *The American Ideology of National Science, 1919–1930*. Pittsburgh: University of Pittsburgh Press.

Tocqueville, Alexis de. 1876. *Democracy in America*. Vol. 2. Boston: John Allyn.

Uchitelle, Louis. 1989. "U.S. Companies Lift R&D Abroad." *New York Times*, 22 February, D2.

U.S. Congress. Office of Technology Assessment. 1985. *Commercial Biotechnology: An International Analysis*. Washington, D.C.: U.S. Government Printing Office.

————. 1993. *Defense Conversion: Redirecting R&D. OTA-ITE-552*. Washington, D.C.: U.S. Government Printing Office.

Usselman, Steven W. 1985. "Running the Machine: The Management of Technological Innovation on American Railroads, 1860–1910." Ph.D. diss., University of Delaware.

————. 1991. "Patents Purloined: Railroads, Inventors, and the Diffusion of Innovation in 19th-Century America." *Technology and Culture* 32: 1047–75.

————. 1992. "From Novelty to Utility: George Westinghouse and the Business of Innovation during the Age of Edison." *Business History Review* 66: 251–304.

Uttal, Bro. 1983. "The Lab That Ran Away From Xerox." *Fortune* 108 (5 September): 97–102.

Vagtborg, Harold. 1975. *Research and American Industrial Development: A Bicentennial Look at the Contributions of Applied R&D*. New York: Pergamon Press.

Van Dyke, Vernon. 1964. *Pride and Power: The Rationale of the Space Program*. Urbana: University of Illinois Press.

Veysey, Laurence. 1965. *The Emergence of the American University*. Chicago: University of Chicago Press.

Vincenti, Walter G. 1990. *What Engineers Know and How They Know It*. Baltimore: Johns Hopkins University Press.

von Braun, Christoph-Friedrich. 1990. "The Acceleration Trap." *Sloan Management Review* 32, no. 1: 49–58.

von Hippel, Eric. 1983. "Novel Product Concepts from Lead Users: Segmenting Users by Experience." Working paper 1476-83. Sloan School of Management, Massachusetts Institute of Technology.

———. 1988. *The Sources of Innovation*. New York: Oxford University Press.

Wall Street Journal. 1991. "U.S.'s DNA Patent Moves Upset Industry," 22 October, B4.

Ward, Patricia S. 1981. "The American Reception to Salvarsan." *Journal of the History of Medicine* 36: 44–62.

Weart, Spencer. 1979. "The Physics Business in America, 1919–1940: A Statistical Reconnaissance." In *The Sciences in the American Context: New Perspectives*, edited by Nathan Reingold. Washington, D.C.: Smithsonian Institution Press.

Weiner, Charles. 1970. "Physics in the Great Depression." *Physics Today* 23 (October): 31–8.

———. 1973. "How the Transistor Emerged." *IEEE Spectrum* 10 (January): 24–33.

Werner, Jerry. 1992. "Technology Transfer in Consortia." *Research-Technology Management* 35, no. 3: 38–43.

Wheelwright, Steven C., and Kim B. Clark. 1992. *Revolutionizing Product Development*. New York: The Free Press.

Whitehead, Alfred North. 1985. Reprint. *Science and the Modern World*. London: Free Association Books. (Original edition, London: The Macmillan Company, 1926).

Wiener, Norbert. 1993. *Invention: The Care and Feeding of Ideas*. Cambridge: MIT Press.

Wilson, Edith. 1990. Product Definition Factors for Successful Designs. Master's thesis, Stanford University.

Wise, George. 1984. "Science at General Electric." *Physics Today* 37, no. 12: 52–61.

———. 1985a. "R&D at General Electric, 1878–1985." Paper presented at the R&D Pioneers Conference, 7 October, Wilmington, Delaware.

———. 1985b. *Willis R. Whitney, General Electric, and the Origins of U.S. Industrial Research*. New York: Columbia University Press.

Wright, Susan. 1986. "Recombinant DNA Technology and Its Social Transformation, 1972–1982." *Osiris*, 2d ser., no. 2: 303–60.

Yerkes, Robert, ed. 1920. *The New World of Science: Its Development During the War.* New York: Century.

INDEX

ABOUT THE CONTRIBUTORS

John Armstrong spent 30 years in IBM's R&D organization. Before his retirement in 1993, he was the director of research and vice president of science and technology.

He is a member of the National Academy of Engineering, a foreign member of the Royal Swedish Academy of Engineering Sciences, a trustee of Associated Universities, Inc., and a member of the Harvard University Board of Overseers. In 1993–94, he was the Karl T. Compton Visiting Lecturer at MIT, and was elected in 1995 to the Council of the National Academy of Engineering. Mr. Armstrong is currently a director of Advanced Technology Materials, Inc., in Danbury, Connecticut, and a visiting professor in the MIT physics department.

Peter R. Bridenbaugh is an executive vice president and the chief technical officer for Alcoa, where he has worked since 1958. He currently heads Alcoa's automotive efforts.

Dr. Bridenbaugh serves on advisory boards for many educational institutions including Carnegie-Mellon, Penn State, Stanford, MIT, Lehigh, Northwestern, and also Carnegie-Bosch. He is a member of the board of directors for Precision Castparts Corporation, the Penn State Research Foundation, and the Alcoa Foundation. In addition, he is a member of the National Academy of Engineers, Sigma Xi, AIME, ASM, the Directors of Industrial Research, and the Industrial Research Institute. Dr. Bridenbaugh is a fellow of ASM International, has received the TMS Leadership Award, and, most recently, is the recipient of the National Materials Advancement Award.

John L. Doyle is a consultant in business planning, technical strategy, and implementation and R&D efficiency. He specializes in electronic and integrated circuit businesses. In these fields, his clients include Hughes, KLA, SVG, and ADI. He previously worked for Hewlett-Packard for thirty-four years, most recently as executive vice president, and had top management responsibility for diverse areas, including Personnel, R&D, and HP Labs, the computer and peripherals businesses, corporate development, engineering, and integrated circuit manufacturing.

Mr. Doyle is a director of Analog Devices, Xilinx, and several private start-up companies, and is a member of the President's Cabinet of

California Polytechnic State University, the President's Council of the Coyote Point Museum, and a Governor of Christ's Hospital.

David A. Hounshell is the Henry R. Luce Professor of Technology and Social Change at Carnegie Mellon University. In 1987–1988, he was a Marvin Bower Fellow at Harvard Business School. He is the author of *From the American System to Mass Production, 1800–1932: The Development of Manufacturing Technology in the United States* and coauthor, with John Kenly Smith, Jr., of *Science and Corporate Strategy: Du Pont R&D 1902–1980*. Recently, Professor Hounshell has conducted research on the development of automation in the post-World War II American automobile industry and is presently carrying out a study of the history of the RAND Corporation in Santa Monica, California.

Dorothy Leonard-Barton is the William J. Abernathy Professor of Business Administration at the Harvard Business School, where she teaches classes on new product and process development, manufacturing, technology strategy, and technology implementation. She is also a faculty member of Harvard programs on Enhancing Corporate Creativity and Managing International Collaboration.

Professor Leonard-Barton's major research interests and consulting expertise are in technology development and commercialization. She is currently undertaking an international study of these topics in the multimedia industry, with special emphasis on managing geographically dispersed research and development teams. Her most recent book, *Wellsprings of Knowledge: Building and Sustaining the Sources of Innovation*, was published by Harvard Business School Press in 1995.

Gordon E. Moore is currently chairman of the board of Intel Corporation, which he co-founded in 1968. Previously, Mr. Moore was director of R&D for the Fairchild Semiconductor Division of Fairchild Camera and Instrument Corporation, a company he also co-founded. He is a director of Varian Associates and Transamerica Corporation, a member of the National Academy of Engineering, a fellow of the IEEE and chairman of the trustees of the California Institute of Technology. Mr. Moore received the National Medal of Technology in 1990.

David C. Mowery is a professor of business and public policy at the Walter A. Haas School of Business at University of California, Berkeley, a research associate at the Canadian Institute for Advanced Research, and deputy director of the Consortium on Competitiveness and Cooperation, a multi-university research alliance dedicated to research on technology management and U.S. competitiveness. Professor Mowery's research focuses on the economics of technological innovation and the effects of public policies on innovation. He has served as an adviser for the Organization for Economic Cooperation and Development, and various federal agencies and industrial firms. Professor Mowery has published numerous academic papers, and has written and edited a number of books.

Mark B. Myers is a senior vice president of corporate research and technology at Xerox Corporation, where he has worked since 1964. Dr. Myers research interests involve digital imaging systems and the creation of new business enterprises involving emerging technology. He serves on several advisory boards with an interest in science and engineering education and government technology and economic policy. Dr. Myers has held visiting professor positions at the University of Rochester and Stanford University.

Richard R. Nelson is the George Blumenthal Professor of International and Public Affairs, Business and Law at Columbia University. His research has largely focused on the process of long run economic change, with particular emphasis on technological advance and on the evolution of economic institutions. Dr. Nelson has written many articles and books on technology and economic growth. His book, *The Sources of Economic Growth*, is forthcoming from Harvard University Press.

Nathan Rosenberg is the Fairleigh S. Dickinson, Jr. Professor of Public Policy in the Department of Economics at Stanford University. His research interests are in the economics of technological change and the economics of science. He is the author of numerous articles and books on these subjects. His most recent book is *Exploring the Black Box* (Cambridge University Press, 1994).

Richard S. Rosenbloom is the David Sarnoff Professor of Business Administration at Harvard Business School, where he has taught courses on competition and strategy, the management of technology, manufacturing policy, and operations management. Professor Rosenbloom serves as a member of the board of directors of Lex Service PLC, Arrow Electronics Corporation, and Executone Information Systems, Inc. He has written articles on technology and innovation for journals such as *Harvard Business Review, California Management Review, Industrial and Corporate Change,* and *Business History Review,* and was the founding editor of the series *Research on Technological Innovation, Management and Policy* (JAI Press).

William J. Spencer is the president and CEO of SEMATECH, a research and development consortium in the semi-conductor industry. He has held key research positions at Xerox Corporation, Sandia National Laboratories, and Bell Telephone Laboratories. Before accepting his current position at SEMATECH in 1990, Dr. Spencer was Group Vice President and Senior Technical Officer at Xerox Corporation in Stamford, Connecticut. He is also research professor of medicine at the University of New Mexico School of Medicine.

Dr. Spencer was awarded the Regents Meritorious Service Medal from the University of New Mexico in 1981, and an honorary doctorate degree from William Jewell College in 1990. He is a member of the National Academy of Engineering, a Fellow of IEEE, and serves on the boards of Adobe Systems and SRI International. Dr. Spencer has written numerous articles for journals such as *California Management Review, Technology Strategy,* and *Medical Progress through Technology.*

David J. Teece is the Mitsubishi Bank Professor at the Haas School of Business and director of the Institute of Management, Innovation, and Organization at UC Berkeley. He is also director of the Consortium on Competitiveness and Cooperation, and co-director of the Management of Technology program. His interests lie in industrial organization and the economics of technological change. He is an expert in technology transfer and the management of innovation and has co-authored over 100 publications. Professor Teece is a co-editor with Richard Rumelt and Dan Schendel, of *Fundamental Issues in Strategy* (Harvard Business School Press).